MathWise Algebra
Book 1
With Answer Key

Peter Wise

MATH TEACHER,
MONUMENT,
COLORADO

CONTRIBUTORS

David Wise

Katherine Wise

Cover Design by Elizabeth Novak

Dedicated to the students in my after-school math clubs, whose commitment and interest in math motivated me to write this workbook.

…with special thanks to:

- *Aileen Finnegan, principal, for her support and encouragement of the math clubs at our school*
- *Shi Hayes, school office manager, whose comments and suggestions were always insightful*
- *My wife, Allison, who assisted at every juncture with helpful comments and opinions*

MathWise Algebra, Book 1 with Answer Key

MathWise Curriculum Press

First printing 2016

MathWise Algebra, Book

TABLE OF CONTENTS

HOW TO USE THIS BOOK

While students can use this series profitably when working alone, my experience has indicated that they will make greater progress if guided by a parent, tutor, or helper. This is particularly true for younger students. This person need not be a math teacher at all—just a reader.

If a student or parent is unclear about a solution or procedure, he or she is encouraged to check with the answer key at the back of the book.

Web Site
For questions, more information, or assistance in starting an algebra club visit www.mathwisebooks.com.

Several years ago I came up with the idea of leading an after-school Algebra Club at the school where I teach in Monument, Colorado. The club met once a week for 24 weeks during the school year. The presentations and materials had to be upbeat and motivating to sustain interest throughout the year. I noticed early on that the students preferred my hand-drawn pencil-on-paper worksheets (done in the style of this book) over the sterile-looking worksheets from most of the resources I could find.

Everyone within a wide range of levels was able to understand the basics of algebra with enough practice. Students often commented that they liked instruction that was out of the ordinary and even wacky. They also found that tips and tricks helped to make the concepts more memorable.

In time, I developed a routine for the Algebra Club, integrating presentations, movie clips, document camera instruction, number drills, and curriculum studies (not to mention snacks). Many of the pages in this book are based on the worksheets we did in this club. As time went on, the numbers increased progressively, until I had 78 students in four different after-school clubs—in one of them I taught highlights of Algebra II and Precalculus concepts to sixth-grade students.

The effects were noticeable. Most of the students in the clubs grew in their understanding of both lower and higher math at an accelerated pace.

I have often told parents that algebra teaches students universals of mathematical concepts irrespective of numbers. It also fosters abstract thinking and clear logical thought. To my surprise, however, I discovered that solving basic algebra problems also helps students with their understanding of lower math (especially basic number facts). Rather than practicing addition with problems like $24 + 9$, it is better for students in upper grades to practice with problems that incorporate algebra ($24x + 9x$, etc.).

This book can be used as a resource for such an enrichment class. With younger students, I would consider beginning first with MathWise Integers; with older students I might use this book first, then move on to MathWise Linear Equations.

It is my hope that this book (and this series) will cultivate solid mathematical understanding among students, and that the math will be made as enjoyable as possible along the way!

Peter Wise
Monument, Colorado

This series of books is designed to be unique and to catch kids' attention in special ways:

Tips and Tricks
Over the years, I have assembled a wide assortment of memory aids—my tips and tricks. Students have found these to be helpful and memorable, but they have also found that these pointers add interest and excitement to their math studies.

Speech Bubbles with Teacher Insights
Speech bubbles are used to provide guidance, point out insights, or give helpful hints as students are solving math problems. Students learn best by doing, and the instruction given in the speech bubbles is designed to (1) sharpen students' powers of observation, (2) increase number sense, and (3) instruct in digestible chunks.

Simplicity of Instruction
Concepts are explained clearly and simply. Every page (excluding review pages or quizzes) has a specific focus. Most pages have generous amounts of white space to keep students focused. Movement is from the simple to the increasingly complex.

Step-By-Step Procedures
Students learn best when given explicit, step-by-step instruction. When several steps are involved, they are numbered. This makes learning much more logical and memorable.

Depth and Complexity
Throughout the book there are challenge problems to stretch students' thinking. At your discretion, you can guide students through the more challenging problems (recommended) or, alternatively, you can have them skip these harder problems.

Informal Terms
This book often employs informal language like "top number" or "bottom number" to keep things simple and focused. Standard mathematical terminology, such as numerator and denominator, is used after the concepts are presented.

Logical-Sequential Instruction
Math problems are presented in a logical sequence, so that previous problems contribute to students' abilities to solve future problems. The order in which you present math problems is critical to promoting number sense.

Basic Algebra Terms

STUDY THESE ANSWERS AND THEN DO THE NEXT PAGE

1. a, x, y

Letters in algebra are called:

variables

because

their values vary

2. We usually don't use x for multiplication in higher math because

x as a letter can be confused with x for multiplication

3. There are three ways to show multiplication:

THIS SYMBOL IS USED MOSTLY IN LOWER GRADES!

A. **x, as in 3 x 4**

B. **dot, as in 3 • 4**

C. **parentheses: 3(4), (3)4, or (3)(4)**

4. Rule: "When they touch…"

"they times" (multiply)

5. $(5)x = 15$

When a number is touching (multiplying) a letter it is called a

WOW! THAT'S A BIG WORD! AM I IN COLLEGE ALREADY?

coefficient

6. You can write 3 times 4 three different ways using parentheses:

3(4)   **(3)4**

(3)(4)

7. You use a letter (doesn't matter which) when you see these words in a problem:

"a number"

"some number"

8. Use an equal sign when you see these words:

"is"   **"the result is"**

Basic Algebra Terms

1. a, x, y

Letters in algebra are called:

[ ]

because

[ ]

2. We usually don't use x for multiplication in higher math because

[ ]

3. There are three ways to show multiplication:

A. [ ]

THIS SYMBOL IS USED MOSTLY IN LOWER GRADES!

B. [ ]

C. [ ]

4. Rule: "When they touch... [ ]

5. ⑤x = 15

↑
When a number is touching (multiplying) a letter it is called a

[ ]

WOW! THAT'S A BIG WORD! AM I IN COLLEGE ALREADY?

6. You can write 3 times 4 three different ways using parentheses:

[ ] [ ]

[ ]

7. You use a letter (doesn't matter which) when you see these words in a problem:

[ ]

8. Use an equal sign when you see these words:

[ ] [ ]

Introducing Variables

- Variables are letters, for example: "a" or "x" or "y"

- They stand for numbers ("mystery numbers" that you have to figure out)

- In different problems, variables represent different numbers

- The numbers VARY (that's why they are called VARIables)

Examples

You may have seen math problems like...

$$7 + \boxed{} = 10 \qquad \boxed{} = 3$$

$$7 + \ ? \ = 10 \qquad ? \ = 3$$

IN THE SAME WAY YOU CAN HAVE...

$$7 + \ x \ = 10 \qquad x \ = 3$$

A BOX, A QUESTION MARK, OR A LETTER ALL HAVE THE SAME IDEA, BUT IN ALGEBRA WE USUALLY USE LETTERS!

Find the number that the variable represents

1. $x + 2 = 8$ $x = \boxed{}$

2. $3 + y = 7$ $y = \boxed{}$

3. $5 + c = 12$ $c = \boxed{}$

4. $a + 6 = 14$ $a = \boxed{}$

5. $10 - x = 8$ $x = \boxed{}$

6. $16 - a = 11$ $a = \boxed{}$

7. $y - 3 = 6$ $y = \boxed{}$

8. $7 + b = 15$ $b = \boxed{}$

9. $14 + x = 20$ $c = \boxed{}$

10. $40 - d = 31$ $d = \boxed{}$

11. $18 - x = 15$ $x = \boxed{}$

12. $a + a = 16$ $a = \boxed{}$

3

Introducing Variables: Multiplication

• Variables, like numbers, can be added, subtracted, multiplied, or divided

Find the number that the variable represents (multiplication)

1. $3 \cdot y = 12$ y = ☐

   A DOT IS USED IN ALGEBRA, BECAUSE "X" IS USUALLY USED AS A VARIABLE!

2. $x \cdot 2 = 14$ x = ☐

3. $4 \cdot n = 44$ n = ☐

4. $9 \cdot r = 27$ r = ☐

5. $a \cdot 2 = 24$ a = ☐

6. $8 \cdot y = 56$ y = ☐

7. $6 \cdot c = 54$ c = ☐

8. $x \cdot 5 = 35$ x = ☐

9. $m \cdot 3 = 24$ m = ☐

10. $a \cdot a = 36$ a = ☐

   SAME LETTERS = SAME NUMBERS

Find the number that the variable represents—WATCH THE SIGNS!

11. $a + 3 = 12$ a = ☐

12. $b \cdot 3 = 33$ b = ☐

13. $4 + r = 10$ r = ☐

14. $e - 2 = 8$ e = ☐

15. $y + 10 = 23$ y = ☐

16. $d \cdot 7 = 42$ d = ☐

4

Introducing Coefficients

Examples

- Coefficients are numbers right before letters—like 2a
- The rule is "if they touch, they times!"
- Coefficients MULTIPLY the letters next to them

HERE, 2 IS THE COEFFICIENT AND A IS THE VARIABLE

$\boxed{2a}$ = 2 times a

(2 · a)

$\boxed{ab}$ = a times b

(a · b)

Use the examples above to solve the following problems

1. 5a = ☐ • a

2. 3y = ☐ • y

3. 7x = ☐ • x

4. 10n = ☐ • n

5. 14r = ☐ • r

6. yz = ☐ • ☐

LITTLE HARDER...

7. 3 · y = ☐

8. 12 · s = ☐

NOW WRITE THESE AS THE NUMBER AND TERM TOUCHING!

9. $\frac{1}{2}$ · a = ☐

10. x · y = ☐

LITTLE HARDER...

11. a · b · c = ☐

12. fgh = ☐ • ☐ • ☐

13. 2mn = ☐ • ☐ • ☐

14. $\frac{3}{4}$ rs = ☐ • ☐ • ☐

5

Coefficients

© Peter Wise, 2014

Examples

- Remember! Coefficients MULTIPLY the LETTERS next to them

- Multiplication is REPEATED ADDITION

$$5 + 5 + 5 = 3 \cdot 5$$

$$y + y + y = 3 \cdot y = 3y$$

YOU ADD THREE Y'S!

... SO IT'S 'THREE TIMES Y!'

Use the examples above to solve the following problems

1. $a + a = \boxed{} a$ HOW MANY A'S ARE BEING ADDED?

2. $s + s + s = \boxed{} s$

3. $4n = \boxed{ + + }$

4. $x + x + x + x + x = \boxed{}$

5. $m + m = \boxed{}$

   $m + m + m = \boxed{}$

   How many m's total (in both lines)? $\boxed{}$

6. $4c = \boxed{ + + }$

   $2c = \boxed{ + }$

   How many c's are being added? $\boxed{} c$

PUTTING IT ALL TOGETHER...

11. $(n + n + n) + (n + n + n + n) =$

    $\boxed{} n + \boxed{} n = \boxed{} n$ HOW MANY N'S TOTAL ARE BEING ADDED?

6

Add the Numbers, Copy the Letters

© Peter Wise, 2014

Examples

A. $2a + 3a = 5a$

THE LETTER STAYS THE SAME!

The numbers touching letters are the **coefficients**

RULE:
Add the numbers (coefficients)
Copy the letters

B.
$$
\begin{array}{r}
4a \\
+\ 3a \\
\hline
7a
\end{array}
$$

YOU CAN ALSO ADD THESE VERTICALLY!

Add or subtract the following variables

1. $3a + 4a = \boxed{}\ a$

2. $5y + 5y + 6y = \boxed{}\ y$

3. $12n + 7n = \boxed{}$

4. $50r + 20r + 6r = \boxed{}$

5. $100x + 100x + 30x + 40x = \boxed{}$

6. $600a + 700a + 8a + 40a = \boxed{}$

7. $30y + 4000y + 50y + 2y = \boxed{}$

8.
$$
\begin{array}{r}
5x \\
+\ 2x \\
\hline
\boxed{}\ x
\end{array}
$$

9.
$$
\begin{array}{r}
37a \\
+\ 45a \\
\hline
\boxed{}
\end{array}
$$

10.
$$
\begin{array}{r}
389m \\
+\ 643m \\
\hline
\boxed{}
\end{array}
$$

Subtract the Numbers, Copy the Letters

Examples

A. $8a - 2a = 6a$

THE LETTER STAYS THE SAME!

The numbers touching letters are called **coefficients**

RULE:

Subtract the numbers
Copy the letters

B.
$$\begin{array}{r} 9x \\ -\ 5x \\ \hline 4x \end{array}$$

YOU CAN ALSO SUBTRACT THESE VERTICALLY!

Add or subtract the following variables

1. $20x - 3x = \boxed{}\,x$

2. $100y - 10y = \boxed{}\,y$

3. $12a - 4a = \boxed{}$

4. $17n - 3n = \boxed{}$

5. $43z - 4z = \boxed{}$

6. $10r + 8r - 3r = \boxed{}$

HERE YOU BOTH ADD AND SUBTRACT THE NUMBERS!

7. $20y + 20y - 2y = \boxed{}$

8. $15c + 15c - 6c = \boxed{}$

9. $28a + 10a - 1a = \boxed{}$

10. $14m + 14m - 6m = \boxed{}$

11.
$$\begin{array}{r} 35a \\ -\ 17a \\ \hline \end{array}$$
$\boxed{}\,a$

12.
$$\begin{array}{r} 63n \\ -\ 25n \\ \hline \end{array}$$
$\boxed{}$

8

Adding and Subtracting Variables

A. $2x + 3x$

The numbers touching letters are called [ ]

You add or subtract the coefficients, but keep the variables the same.

$2x + 3x = 5x$

| $x + x$ | $x + x + x$ | $x + x + x$ $x + x$ |

add just the numbers: $2 + 3 = 5$

$2x + 3x = 5x$

the variable stays the same

Add or subtract the following variables

1. $4x + 3x =$ [ ]

2. $12x - 4x =$ [ ]

3. $5x + 14x =$ [ ]

4. $6a - 2a =$ [ ]

5. $10y + 2y + 4y =$ [ ]

6. $12r - 5r + 20s - 4s =$

[ ] r + [ ] s

LETTERS BY THEMSELVES

Letters that are by themselves have an invisible 1 in front of them: $y = 1y$

$y + y = 2y$
$1y + 1y = 2y$

THESE ARE THE SAME!

7. $15x + x =$ [ ] x

8. $8a - a + 3b + b =$ [ ] a + [ ] b

9

The Invisible ONE

A. $a + a = 2a$

1a + 1a

NO NUMBER IN FRONT OF A LETTER? THEN AN INVISIBLE ONE IS ACTUALLY IN FRONT OF IT!

Any time you see a letter by itself, then there is really an invisible one in front of it

Add or subtract the following terms

1. $y + y + y = \boxed{}\, y$

2. $6n - n = \boxed{}$

3. $12x + 2x - x = \boxed{}$

4. $590a + 10a - a = \boxed{}$

5. $r + r + 5r = \boxed{}$

6. $4m + m + 3m = \boxed{}$

7. $12z + z + 2z = \boxed{}$

8. $20a - a + 20a - a = \boxed{}$

9. $15y + y + 15y + y = \boxed{}$

10. $\begin{array}{r} 15n \\ -\ n \\ \hline \boxed{} \end{array}$

11. Rewrite the following problem with coefficients (numbers in front of letters)

$(n + n + n) - n - n$

$\boxed{} - \boxed{} = \boxed{}$

NOW TRY THESE PROBLEMS!

12. $a + 3000a + 20a + 700a = \boxed{}$

13. $500x + 1200x + 28x - x = \boxed{}$

Adding Variables with Exponents

Examples

A. $3\boxed{a^2} + 4\boxed{a^2} = 7\boxed{a^2}$

Add the numbers
Copy the letters and exponents

RULE:

You can only add or subtract terms if they have the
* Same letter AND
* Same exponent

B.
$$+\begin{array}{r} 6x^3 \\ 4x^3 \\ \hline 10x^3 \end{array}$$

Add or subtract the following terms

AFTER YOU ADD THE COEFFICIENTS, JUST COPY ALL THE LETTERS AND EXPONENTS THEY WAY THEY ARE IN BOTH TERMS!

1. $4x^3 + 5x^3 = \boxed{}\,x^3$

2. $2y^9 + 6y^9 = \boxed{}\,\boxed{}^{\boxed{}}$

3. $2n^2 + 2n^2 + 2n^2 = \boxed{}\,\boxed{}^{\boxed{}}$

4. $3a^2 + 4a^2 - 2a^2 = \boxed{}\,a^2$

SUBTRACTION WORKS THE SAME WAY AS ADDITION!

5. $10x^2 + 2x^2 - 5x^2 = \boxed{}\,\boxed{}^{\boxed{}}$

6. $10a^2\,b^3 + 5a^2\,b^3$

$= \boxed{}\,\boxed{}^{\boxed{}}\,\boxed{}^{\boxed{}}$

7. $2x^5\,y^2\,z^4 + 3x^5\,y^2\,z^4$

$= \boxed{}\,\boxed{}^{\boxed{}}\,\boxed{}^{\boxed{}}\,\boxed{}^{\boxed{}}$

8. $16a^7\,b^3 - 2a^7\,b^3 - 4a^7\,b^3$

$= \boxed{}\,\boxed{}^{\boxed{}}\,\boxed{}^{\boxed{}}$

11

Intro to Substitution

Example

USE PARENTHESES WHERE EACH VARIABLE IS, THEN SUBSTITUTE THE NUMBER FOR THE VARIABLE AND CALCULATE!

A. ab $a = 2$

$a \cdot b$ $b = 5$

$(2)(5) = 10$

↑ ↑

substitute 2 for the letter a   substitute 5 for the letter b

---

Substitute values for the variables and calculate

1. xy $x = 4$ $y = 3$

$(\quad)(\quad) =$ ☐

REMEMBER! WHEN THEY TOUCH, THEY TIMES!

2. ab^2 $a = 2$ $b = 3$

$(\quad)(\quad^2) =$ ☐

3. a^2b $a = 5$ $b = 2$

$(\quad^2)(\quad) =$ ☐

4. a^2b^2 $a = 3$ $b = 10$

$(\quad)(\quad) =$ ☐

write the exponents inside the parentheses

5. $\dfrac{a}{2} \cdot b$ $a = 8$ $b = 10$

$\dfrac{(\quad)}{2}(\quad) =$ ☐

Order of Operations: multiply before you add

6. $xy + z$ $x = 4$ $y = 8$ $z = 2$

$=$ ☐

7. $\dfrac{m + n}{3}$ $m = 20$ $n = 7$

$\dfrac{(\quad) + (\quad)}{3} =$ ☐

8. $r - st$ $r = 20$ $s = 4$ $t = 3$

☐ $=$ ☐

Substitution Practice

1. $ab - cd$ $a = 4$ $b = 10$
$c = 3$ $d = 5$

TRADE THE LETTER A FOR 4!   TRADE THE LETTER B FOR 10!

$($ $)($ $) - ($ $)($ $)$
$a = 4$ $b = 10$ $c = 3$ $d = 5$

$\boxed{} - \boxed{} = \boxed{}$

2. πr^2 $\pi = 3.14$ $r = 3$

$($ $)($ $)^2 = \boxed{}$

Note: π is not a variable, it is a constant, meaning that it will always have a definite value; we'll use 3.14

3. $\dfrac{n^2}{2}$ $n = 4$

$\dfrac{\boxed{}}{2} = \boxed{}$

4. $3y^2$ $y = 10$

$($ $)($ $) = \boxed{}$

Note:
Only the y is raised to the second power

5. $4ac$ $a = 3$ $c = 4$

PUT EVERYTHING IN SEPARATE PARENTHESES!

$= \boxed{}$

6. $b^2 - 4ac$ $a = 2$ $b = 8$
$c = 3$

$= \boxed{}$

7. $\dfrac{a^2}{b} \cdot c$ $a = 6$ $b = 4$
$c = 3$

$= \boxed{}$

8. $a + bc$ $a = 12$ $b = 7$
$c = 8$

Watch for the order of operations!

$= \boxed{}$

9. $5x^2 - xy$ $x = 3$ $y = 11$

$= \boxed{}$

10. $r^2st - (r + s)$ $r = 3$ $s = 2$ $t = 4$

$= \boxed{}$

13

Turning Words into Algebra Symbols

Key words or phrases to look for:

"a number" "some number"	→	x (or any other variable letter)

"is" "the result is"	→	= (equal sign)

"sum" → addition

"increased by" → addition

"difference" → subtraction

"decreased by" → subtraction

"product" → multiplication

"quotient" → division

Watch out for these

"3 less than a number" → $x - 3$

watch the order!   variable comes first

"3 more than a number" → $x + 3$

Translate the following words into algebraic symbols; solve them if you can!

Examples

A.	The sum of 5 and a number is 25	$5 + x = 25$	x =
B.	The product of 4 and a number is 12	$4x = 12$	x =

1.	A number increased by 10 is 40		x =
2.	2 less than a number is 5		x =
3.	The quotient of a number and 2 is 16		x =
4.	The difference of a number and 6 is 10		x =

14

Identify the Equation

1. A number is increased by 3; the result is 20

   (a) $x + 3 = 20$

   (b) $3x = 20$

2. The product of a number and 8 is 3 less than 35

   (a) $8x = 3 - 35$

   (b) $8x = 35 - 3$

3. The product of a number and 5 equals the sum of the number and 16

   (a) $5x = x + 16$

   (b) $5x + 16 = x$

4. The difference of 40 and a number equals the product of the (same) number and 3

   (a) $x - 40 = 3x$

   (b) $40 - x = 3x$

5. The quotient of a number and 6 equals the (same) number decreased by 20

   (a) $x \div 6 = 20 - x$

   (b) $x \div 6 = x - 20$

6. 6 less than a number equals the quotient of the number and 2

   (a) $x - 6 = x \div 2$

   (b) $6 - x = x \div 2$

7. The sum of a number and 10 equals the difference of 20 and 2

   (a) $x + 10 = 20 - 2$

   (b) $x + 10 = 20 \div 2$

8. The product of a number and 3 is increased by 8. The result is 20.

   (a) $3x \cdot 8 = 20$

   (b) $3x + 8 = 20$

9. 5 less than a number is equal to the quotient of 30 and 3

   (a) $5 - x = 30 \div 3$

   (b) $x - 5 = 30 \div 3$

© Peter Wise, 2014

More Words into Algebra Symbols

More Key words or phrases to look for:

"doubled"	$\longrightarrow$	multiplied by 2
"tripled"	$\longrightarrow$	multiplied by 3
"halved"	$\longrightarrow$	divided by 2
"squared"	$\longrightarrow$	raised to the 2nd power
"cubed"	$\longrightarrow$	raised to the 3rd power

Circle (a) or (b) to identify the equation that correctly represents the words

1. A number is doubled. The result is equal to half of 24.

    (a) $2x = 24 \cdot 2$
    (b) $2x = 24 \div 2$

2. A number squared equals the sum of the (same) number and 30

    (a) $x^2 = x + 30$
    (b) $x^2 + x = 30$

3. Triple a number equals the (same) number squared, then decreased by the (same) number

    (a) $x^3 = x - x^2$
    (b) $3x = x^2 - x$

4. A number cubed, then increased by 3 equals the product of the number and 10

    (a) $x^3 + 3 = 10x$
    (b) $x^3 + 10 = 3x$

5. The difference of a 20 and a number is equal to the (same) number squared

    (a) $20 - x = x^2$
    (b) $20 - x = 2x$

6. Half of a number is equal to 2 less than 20

    (a) $x \div 2 = 20 - 2$
    (b) $x \div 2 = 20 \div 2$

Turning Words into Algebra Symbols

Translate the following words into algebraic symbols

You don't need to solve these equations;
just write them for now

1. Triple a number is 15

2. Double a number; decrease it by 4; the result is 20

3. 5 less than a number is 30

watch the order on these! ↑ ↓

4. 5 more than a number is 45

5. A number is increased by 4, then squared, the result is 144

   HINT!
   YOU'LL NEED TO USE
   PARENTHESES!

6. A number squared, then increased by 3, is 12

7. A number squared equals 5 times the number, decreased by 4
   (same variable)

8. A number tripled equals the number increased by 14

9. Some number is doubled; it equals 24 decreased by that number

10. A number doubled, then increased by 2 equals the number tripled, then decreased by 7

17

Like Terms

- Matching Letters
- Matching Exponents (on the letters)

IT DOESN'T MATTER WHAT THE COEFFICIENTS ARE!

Examples

A. $2x \quad x$

These are like terms

$x^3 \quad 2x^2$

These are NOT like terms

B. These are all like terms

$3x^5y^2 \quad 3y^2x^5 \quad 2yyx^5$

THE ORDER DOESN'T MATTER!

YY IS THE SAME AS Y-SQUARED!

Circle the LIKE TERMS in each row

1. $x \quad 3x \quad x^2$

2. $4x^3 \quad 4x \quad x^3$

3. $5xy \quad 8x^2y^2 \quad 7yx$

4. $7a^3b^4 \quad 2a^5b^6 \quad 9a^5b^6$

5. $2x^2y^3 \quad 3x^3y^2 \quad 6y^2x^3 \quad 2x^3yy$

6. $7^2c^3d^5 \quad 7c^3d^5 \quad 7^2cccd^5$

7. $8x^7y^4z^3 \quad 5x^7y^3z^4 \quad x^7y^3z^4$

8. $3n^3x^2y^4 \quad 4n^5x^2y^3 \quad -(.2)n^5x^2y^3$

Add the LIKE TERMS

Example

A. $3a^3 + 4a^3 + 2a^5 = 7a^3 + 2a^5$

CAN'T BE ADDED WITH THE OTHERS BECAUSE IT'S NOT A LIKE TERM!

1. $5x^2 + 4y^3 + 3x^2 = \boxed{} + \boxed{}$

THE TERM YOU CAN'T ADD PUT HERE!

2. $2a^4b^3c^2 + 8a^5b^4c^3 + 6a^5b^4c^3 = \boxed{} + \boxed{}$

18

Like Terms

Examples

LOOK ONLY AT THE LETTERS AND THEIR EXPONENTS!

COEFFICIENTS DON'T MATTER!

LIKE TERMS:

- Matching letter(s) - order doesn't matter
- Matching exponent(s) - same exponents on the same letters

Like terms can have different coefficients

A. $a^2 \quad a^2$

Like terms

THE LETTERS AND THEIR EXPONENTS MATCH!

B. $x^2y^3 \quad x^2y^3$

Like terms

THE EXPONENT ON EACH X IS THE SAME; THE EXPONENT ON EACH Y IS THE SAME!

C. $3a^4b^5 \quad 7b^5a^4$

Like terms

THE ORDER OF THE LETTERS DOESN'T MATTER!

D. $5n^3m^4 \quad 5n^3$

NOT like terms

M^4 IS MISSING!

E. $2r^2s^7 \quad 2r^3s^7$

NOT like terms

THE EXPONENT ON THIS R IS DIFFERENT FROM THE EXPONENT ON THE OTHER R!

Circle Y or N for Yes or No

1. $2x^3y^4 \quad 3x^3y^4$

a) Same letters? Y N

b) Do the same letters have the same exponents? Y N

c) Are these LIKE TERMS? Y N

3. $5n^7m^9 \quad 5n^8m^9$

a) Same letters? Y N

b) Do the same letters have the same exponents? Y N

c) Are these LIKE TERMS? Y N

2. $3a^4b^5 \quad 7b^5a^4$

ORDER DOESN'T MATTER!

a) Same letters? Y N

b) Do the same letters have the same exponents? Y N

c) Are these LIKE TERMS? Y N

4. $4yyy \quad 9y^3$

$YYY = Y^3$!

a) Same letters? Y N

b) Do the same letters have the same exponents? Y N

c) Are these LIKE TERMS? Y N

19

Like Terms

- You can only ADD or SUBTRACT numbers or variables if they are LIKE TERMS

- If they are NOT like terms, keep them separate (don't add the coefficients)

Examples

A.

LIKE TERMS
You can add these

The exponent on the a doesn't match the other exponents on a
Keep this one separate!

$$\boxed{3a^2 + 4a^2} \quad + 5a^3 =$$

$$\boxed{7a^2} \qquad \boxed{+ 5a^3}$$

YOU CAN'T ADD THE 7 AND THE 5 BECAUSE THE EXPONENTS DON'T MATCH!

Answer: $7a^2 + 5a^3$

Add or subtract the following terms

1. $4x^5 + 6x^5 + 3x^2 = \boxed{} + \boxed{}$

2. $7y^3 + 8y^2 + 2y^3 = \boxed{} + \boxed{}$

LOOK CLOSELY... TERMS THAT CANNOT BE COMBINED WILL HAVE TO BE ADDED OR SUBTRACTED AS THEY ARE!

3. $12n^4 - 2n^4 + 2y^3 = \boxed{} + \boxed{}$

4. $3x^2y^3 + 5x^2y^4 + 4x^2y^3 = \boxed{} + \boxed{}$

Look closely...
One term has to be separate

5. $15r^3s^5 - 2r^2s^2 - 3r^3s^5 = \boxed{} - \boxed{}$

6. $2a^2b^3 + 5x^3y^4 + 6a^2b^3 + 2x^3y^4 = \boxed{} + \boxed{}$

20

Undoing Addition and Subtraction

Goal: Isolate the variable (x = some number)

How to do this:

Undo addition by SUBTRACTION

Undo subtraction by ADDITION

Examples

3 IS BEING ADDED TO THE VARIABLE!

A.
$$a + 3 = 10$$
$$\underline{-3 \quad -3}$$
$$a = 7$$

TO ISOLATE THE VARIABLE, UNDO +3, BY SUBTRACTING 3!

BUT THE RULE IS "WHATEVER YOU DO TO ONE SIDE...YOU HAVE TO DO TO THE OTHER!"

B.
$$a - 2 = 6$$
$$\underline{+2 \quad +2}$$
$$a = 8$$

IN THIS CASE, UNDO MINUS 2 BY ADDING 2 TO BOTH SIDES!

Solve the following equations by reversing the addition/subtraction to the variable

1. $a + 5 = 20$

$- \square \quad - \square$

$a = \square$

THESE SIGNS WILL ALWAYS BE OPPOSITE!

2. $a + 7 = 30$

$- \square \quad - \square$

$a = \square$

3. $n - 5 = 12$

$+ \square \quad + \square$

$n = \square$

4. $6 + m = 15$

$- \square \quad - \square$

$m = \square$

5. $c - 4 = 27$

$+ \square \quad + \square$

$c = \square$

6. $y - 4 = 21$

$+ \square \quad + \square$

$y = \square$

7. $-8 + x = 22$

$+ \square \quad + \square$

$x = \square$

8. $32 = 12 + n$

$- \square \quad - \square$

$\square = n$

DO THIS BOX FIRST!

9. $k - 9 = 34$

$+ \square \quad + \square$

$k = \square$

Undoing Addition and Subtraction

Solve the following equations by reversing the addition/subtraction to the variable

1. $p - 6 = 13$

THESE SIGNS WILL ALWAYS BE OPPOSITE!

$+ \boxed{} \quad + \boxed{}$

$p = \boxed{}$

2. $n - 8 = 28$

$+ \boxed{} \quad + \boxed{}$

$n = \boxed{}$

3. $h + 4 = 32$

$- \boxed{} \quad - \boxed{}$

$h = \boxed{}$

4. $9 + b = 20$

$- \boxed{} \quad - \boxed{}$

$b = \boxed{}$

5. $y - 12 = 36$

$+ \boxed{} \quad + \boxed{}$

$y = \boxed{}$

6. $m - 15 = 46$

$+ \boxed{} \quad + \boxed{}$

$m = \boxed{}$

7. $14 + s = 32$

$- \boxed{} \quad - \boxed{}$

$s = \boxed{}$

8. $v - 10 = 49$

$+ \boxed{} \quad + \boxed{}$

$v = \boxed{}$

9. $x + 18 = 62$

$- \boxed{} \quad - \boxed{}$

$x = \boxed{}$

10. $81 = 17 + a$

$- \boxed{} \quad - \boxed{}$

$\boxed{} = a$

ELIMINATE THE NUMBER ON THE SAME SIDE AS THE VARIABLE!

11. $45 = j - 17$

$+ \boxed{} \quad + \boxed{}$

$j = \boxed{}$

12. $m - 12 = 79$

$+ \boxed{} \quad + \boxed{}$

$m = \boxed{}$

DO THIS BOX FIRST!

13. $41 = 13 + y$

$- \boxed{} \quad - \boxed{}$

$\boxed{} = y$

DO THIS BOX FIRST!

14. $-12 + r = 38$

$+ \boxed{} \quad + \boxed{}$

$r = \boxed{}$

15. $q + 17 = 21$

$- \boxed{} \quad - \boxed{}$

$q = \boxed{}$

Undoing Multiplication & Division

Goal: Isolate the variable (like x = some number)

How to do this:

Undo multiplication by DIVISION

Undo division by MULTIPLICATION

A NUMBER IN THE DENOMINATOR IS JUST ANOTHER WAY TO SHOW DIVISION!

Examples

A.
$$5a = 45$$
$$\boxed{\div 5} \quad \boxed{\div 5}$$
$$a = 9$$

B.
$$a \div 3 = 4$$
$$\boxed{\times 3} \quad \boxed{\times 3}$$
$$a = 12$$

C.
$$\frac{a}{2} = 7$$
$$\boxed{\times 2} \quad \boxed{\times 2}$$
$$a = 14$$

Solve the following equations by reversing the multiplication/division to the variable

WHEN THEY TOUCH, THEY TIMES!

1.
$$3a = 15$$
THESE OPERATIONS WILL ALWAYS BE OPPOSITE!
$$\div \boxed{} \quad \div \boxed{}$$
$$a = \boxed{}$$

2.
$$a \div 7 = 3$$
$$\times \boxed{} \quad \times \boxed{}$$
$$a = \boxed{}$$

3.
$$d \times 6 = 42$$
$$\div \boxed{} \quad \div \boxed{}$$
$$d = \boxed{}$$

4.
$$\frac{a}{6} = 3$$
$$\times \boxed{} \quad \times \boxed{}$$
$$a = \boxed{}$$

5.
$$4s = 48$$
$$\div \boxed{} \quad \div \boxed{}$$
$$s = \boxed{}$$

6.
$$n \div 8 = 7$$
$$\times \boxed{} \quad \times \boxed{}$$
$$n = \boxed{}$$

7.
$$9y = 54$$
$$\div \boxed{} \quad \div \boxed{}$$
$$y = \boxed{}$$

8.
$$24 = 8m$$
$$\div \boxed{} \quad \div \boxed{}$$
$$m = \boxed{}$$

9.
$$\frac{a}{12} = 4$$
$$\times \boxed{} \quad \times \boxed{}$$
$$a = \boxed{}$$

Undoing Multiplication & Division

Solve the following equations by reversing the addition/subtraction to the variable

1. $3a = 33$

$\div \boxed{} \; \div \boxed{}$

$a = \boxed{}$

2. $7y = 28$

$\div \boxed{} \; \div \boxed{}$

$y = \boxed{}$

3. $n \div 8 = 2$

$\times \boxed{} \; \times \boxed{}$

$n = \boxed{}$

4. $\dfrac{m}{5} = 2$

$\times \boxed{} \; \times \boxed{}$

$m = \boxed{}$

5. $\dfrac{y}{9} = 3$

$\times \boxed{} \; \times \boxed{}$

$y = \boxed{}$

6. $\dfrac{x}{3} = 7$

$\times \boxed{} \; \times \boxed{}$

$x = \boxed{}$

7. $48 = 8n$

$\div \boxed{} \; \div \boxed{}$ — DO THIS BOX FIRST!

$n = \boxed{}$

8. $n \div 5 = 7$

$\times \boxed{} \; \times \boxed{}$

$n = \boxed{}$

9. $9r = 54$

$\div \boxed{} \; \div \boxed{}$

$r = \boxed{}$

10. $\dfrac{b}{7} = 7$

$\times \boxed{} \; \times \boxed{}$

$b = \boxed{}$

11. $72 = 6c$

$\div \boxed{} \; \div \boxed{}$

$c = \boxed{}$

12. $n \div 7 = 8$

$\times \boxed{} \; \times \boxed{}$

$n = \boxed{}$

13. $72 = 9x$

$\div \boxed{} \; \div \boxed{}$ — DO THIS BOX FIRST!

$x = \boxed{}$

14. $\dfrac{n}{5} = 8$

$\times \boxed{} \; \times \boxed{}$

$n = \boxed{}$

15. $k \div 12 = 8$

$\times \boxed{} \; \times \boxed{}$

$k = \boxed{}$

24

Do the Algebra Steps

"WHATEVER YOU DO TO ONE SIDE, YOU HAVE TO DO TO THE OTHER!"

(-2) (-2)

1. $3x + 2 = 29$

"CHANGE ONE THING...
COPY AGAIN!"

FIRST, YOU WANT TO
GET RID OF ANYTHING
ADDED OR SUBTRACTED
TO THE VARIABLE!

[ ] = [ ]

÷ [ ]   ÷ [ ]

WHAT DO YOU
DIVIDE BY TO UNDO
"TIMES 3"?

"CHANGE ONE THING...
COPY AGAIN!"

NOW, YOUR
ANSWER:

[=]

(-) (-)

STEP ONE,
GET RID OF
ANYTHING ADDED OR
SUBTRACTED!

2. $4x + 5 = 29$

[ ] = [ ]

THIS IS STEP TWO,
WHEN YOU DIVIDE TO
UNDO THE
MULTIPLICATION!

÷ [ ]   ÷ [ ]

[=]

HERE IS
YOUR
ANSWER!

HOW DO YOU
UNDO MINUS?

STEP ONE:
ANSWER GOES HERE!

3. $6x - 3 = 63$

[ ] = [ ]

STEP TWO:
DIVIDE TO UNDO THE
MULTIPLICATION!

÷ [ ]   ÷ [ ]

[=]   ANSWER!

REMEMBER: THE STEPS ARE JUST
AS IMPORTANT AS THE ANSWER!

4. $7x - 6 = 50$

[ ] = [ ]

÷ [ ]   ÷ [ ]

[=]

5. $8y + 2 = 34$

[ ] = [ ]

÷ [ ]   ÷ [ ]

[=]

6. $4a - 8 = 28$

[ ] = [ ]

÷ [ ]   ÷ [ ]

[=]

LEARN GOOD ALGEBRA
TECHNIQUE! IN THE FUTURE YOU
WILL GET PROBLEMS WITH TOO
MANY STEPS TO DO MENTALLY!

25

Solving 2-Step Equations

1.

YOU WANT THE + 3 TO DISAPPEAR!

$7a + 3 = 17$

"WHATEVER YOU DO TO ONE SIDE, YOU HAVE TO DO TO THE OTHER!"

$- \boxed{3} \quad - \boxed{3}$

$\boxed{7a} = \boxed{14}$

REWRITE THE NEW EQUATION!

$\div \boxed{7} \quad \div \boxed{7}$

UNDO "7 TIMES A"

$\boxed{a = 2}$

2.

YOU WANT THE + 6 TO DISAPPEAR!

$5n + 6 = 21$

$- \boxed{} \quad - \boxed{}$

REWRITE THE NEW EQUATION!

$\boxed{} = \boxed{}$

$\div \boxed{} \quad \div \boxed{}$

UNDO "5 X N"

$\boxed{n = }$

3.

$3y - 2 = 34$

$+ \boxed{} \quad + \boxed{}$

$\boxed{} = \boxed{}$

$\div \boxed{} \quad \div \boxed{}$

$\boxed{y = }$

4.

$6c - 6 = 30$

$+ \boxed{} \quad + \boxed{}$

$\boxed{} = \boxed{}$

$\div \boxed{} \quad \div \boxed{}$

$\boxed{c = }$

5.

$4y + 2 = -26$

$- \boxed{} \quad - \boxed{}$

$\boxed{} = \boxed{}$

$\div \boxed{} \quad \div \boxed{}$

$\boxed{y = }$

6.

$6r + 12 = 60$

$- \boxed{} \quad - \boxed{}$

$\boxed{} = \boxed{}$

$\div \boxed{} \quad \div \boxed{}$

$\boxed{r = }$

Solving 2-Step Equations

1. 3x − 5 = 22

GET RID OF THIS FIRST!

+ ☐   + ☐

☐ = ☐   "CHANGE ONE THING... COPY AGAIN!"

÷ ☐   ÷ ☐

x =

2. 7a + 10 = 52

− ☐   − ☐

☐ = ☐

REWRITE THE NEW EQUATION!

÷ ☐   ÷ ☐

a =

3. 5c + 5 = −45

− ☐   − ☐

☐ = ☐

÷ ☐   ÷ ☐

c =

4. 4n + 8 = 52

− ☐   − ☐

☐ = ☐

÷ ☐   ÷ ☐

n =

5. 8x − 5 = 59

+ ☐   + ☐

☐ = ☐

÷ ☐   ÷ ☐

x =

6. 7y − 9 = 75

+ ☐   + ☐

☐ = ☐

÷ ☐   ÷ ☐

y =

27

Exponents = Number of Repeated Factors

A. x^4

(exponent)

(base)

x is a factor four times

(base) (exponent)

$x \cdot x \cdot x \cdot x$

A factor is a multiplied number

→ | x | is a factor | 4 | times | $x \cdot x \cdot x \cdot x$ |

(base) (exponent) (show the multiplication)

Error alert: It does not mean x times 4

Use the example above as a pattern for the following problems

1. $y^2 \rightarrow$ ☐ is a factor ☐ times ☐

(base) (exponent) (show the multiplication)

2. $a^4 \rightarrow$ ☐ is a factor ☐ times ☐

(base) (exponent) (show the multiplication)

3. $n^3 \rightarrow$ ☐ is a factor ☐ times ☐

(show the multiplication)

4. $r^5 \rightarrow$ ☐ is a factor ☐ times ☐

(show the multiplication)

5. $m^3 \rightarrow$ ☐ is a factor ☐ times ☐

(show the multiplication)

6. Show with an exponent: 3 is a factor 4 times

7. Show with an exponent: 2 is a factor 5 times

(show the multiplication)

8. Show with an exponent: 10 is a factor 3 times

Variables with Exponents

Examples

A. $xx = \boxed{x}^{\boxed{2}}$

2 X'S ARE MULTIPLYING EACH OTHER!

B. $(x + 2)(x + 2) = \boxed{(x + 2)}^{\boxed{2}}$

(X + 2) IS A FACTOR TWO TIMES!

Write the following terms with exponents

1. $yy = \boxed{}^{\boxed{}}$

2. $x^5 = \boxed{\quad\cdot\quad\cdot\quad\cdot\quad\cdot\quad}$

3. $rrr = \boxed{}^{\boxed{}}$

4. $nnnn = \boxed{}^{\boxed{}}$

5. $y^3 = \boxed{}$

6. $mmm = \boxed{}^{\boxed{}}$

7. $d^4 = \boxed{}$

8. $(xyz)(xyz)(xyz) = (xyz)^{\boxed{}}$

9. $(aa)(aa) = (aa)^{\boxed{}} \text{ or } a^{\boxed{}}$

10. $(ab)(ab) = (ab)^{\boxed{}}$

WHEN TERMS INSIDE PARENTHESES ARE IDENTICAL AND MULTIPLIED 1 OR MORE TIMES, THEY CAN ALSO BE WRITTEN WITH EXPONENTS!

11. $(x + 2)(x + 2) = (x + 2)^{\boxed{}}$

12. $(y + 3)(y + 3)(y + 3) = (y + 3)^{\boxed{}}$

13. $(a + 2y)^2 =$

$\boxed{(\quad\quad)(\quad\quad)}$

14. $(2 + m)^3 =$

$\boxed{(\quad\quad)(\quad\quad)(\quad\quad)}$

15. $(xy)^3 =$

$\boxed{}$

Multiplying Numbers and Letters

Examples

A. $a \cdot a = a^2$

(speech bubble) TWO A'S ARE MULTIPLYING EACH OTHER!

(speech bubble) IF NO EXPONENT IS WRITTEN, IT'S AN **INVISIBLE ONE!**

B. $a^1 \cdot a^1 = a^2$

C. $a^2 \cdot a^3 = a^{2+3}$ or a^5

D. $\boxed{3}a \cdot \boxed{2}a = \boxed{6}a^2$ $3 \cdot 2 = 6$

$a \cdot a = $ a squared

E. $\boxed{2}a^2 \cdot \boxed{5}a^4 = \boxed{10}a^6$

Multiply the numbers

Multiply the letters

- Whenever you multiply you add factors (= why you add exponents here)
- You are just adding numbers that are already being multiplied

---

Use the example above as a pattern for the following problems

1. $d \cdot d = d^{\square}$

2. $r^2 \cdot r^2 = r^{\square}$

3. $t^3 \cdot t^7 = t^{\square}$

(speech bubble) IF YOU DON'T SEE AN EXPONENT ON A NUBMER OR A LETTER--IT'S REALLY AN INVISIBLE 1!

4. $m^2 \cdot m^4 \cdot m = m^{\square}$

5. $x^7 \cdot x^7 = x^{\square}$

6. $3x \cdot 4x = \boxed{}x^{\square}$

7. $5y^2 \cdot 2y = \boxed{}y^{\square}$

(speech bubble) HOW MANY Y'S ARE MULTIPLYING EACH OTHER?

8. $6a^3 \cdot 3a^4 = \boxed{}a^{\square}$

9. $2x^{10} \cdot 8x^5 = \boxed{}x^{\square}$

10. $7y^3 \cdot 4y^2 = \boxed{}y^{\square}$

Multiplying Numbers and Letters

Example

A. $3a^2 \cdot 5a^{10} = 15a^{12}$

#1 Multiply the numbers $\qquad 3 \cdot 5 = 15$

#2 Multiply the variables $\qquad a^2 \cdot a^{10} = a^{12}$

EXPONENTS SHOW MULTIPLICATION!

Since the letters and numbers are multiplying each other, put them together ("when they touch, they times!")

$\boxed{15a^{12}}$

Multiply the following terms

1. $2x \cdot 7x = \boxed{} x^{\square}$

2. $2x^3 \cdot 7x^5 = \boxed{} x^{\square}$

3. $4a^5 \cdot 6a^6 = \boxed{} a^{\square}$

4. $8y^2 \cdot 2y^5 = \boxed{} y^{\square}$

5. $12a^7 \cdot 3a^8 = \boxed{} a^{\square}$

6. $3n \cdot 4n^2 = \boxed{} n^{\square}$

7. $5r^3 \cdot 7r^9 = \boxed{} r^{\square}$

8. $3a^2b^3 \cdot 2a^5b^5 = \boxed{} a^{\square} b^{\square}$

9. $6x^5y^6 \cdot 8x^4y^6 = \boxed{} x^{\square} y^{\square}$

10. $-4m^8n \cdot 7m^2n^4 = \boxed{} m^{\square} n^{\square}$

11. $6x^3y^3 \cdot 3x^4y^8 = \boxed{}$

12. $4r^4s^7 \cdot -8r^5s^9 = \boxed{}$

13. $2a^2 \cdot 3a^3 \cdot 5a^4 = \boxed{}$

14. $-6x^4y^3 \cdot -9y^5x^2 = \boxed{}$

WATCH THE ORDER ON THIS ONE!

31

Adding vs. Multiplying Variables

A. $a \cdot a = a^2$

B. $a^2 \cdot a^2 = a^4$

C. $\boxed{3}a^2 \cdot \boxed{2}a^5 = \boxed{6}a^7$

D. $a + a = 2a$

E. $a^3 + a^3 = 2a^3$

F. $3a^5 + 3a^5 = 6a^5$

Keep the variable and exponent the same (when you add them they all have to match)

Add the coefficients (numbers in front of the variable)

Follow the correct pattern to add or multiply the following expressions

1. $x + x = \boxed{}$

2. $(x)(x) = \boxed{}$

When parentheses touch, they times!

3. $3x + 3x = \boxed{}\, x$

4. $3x \cdot 3x = \boxed{}\, x^{\boxed{}}$

Remember to multiply both the numbers and the letters!

5. $4y \cdot 2y^2 = \boxed{}\, y^{\boxed{}}$

6. $5a^3 + 2a^3 = \boxed{}\, a^{\boxed{}}$

7. $5a^3 \cdot 2a^3 = \boxed{}\, a^{\boxed{}}$

8. $2y^5 + 2y^5 + 2y^5 = \boxed{}\,\boxed{}^{\boxed{}}$

HOW MANY y^5s DO YOU HAVE?

9. $2x^7 \cdot 5x^3 \cdot 7x^5 = \boxed{}\,\boxed{}^{\boxed{}}$

10. $(4m^3)(6m^4) = \boxed{}\,\boxed{}^{\boxed{}}$

Dividing Same Bases

© Peter Wise, 2014

Examples

A. $x^5 \div x^2 = x^{5-2} = x^3$

$\dfrac{x^5}{x^2} =$ XXXXX / XX

After you cancel two x's from both top and bottom **3 x's remain**

THE BOTTOM FACTORS SUBTRACT FROM THE TOP FACTORS!

B. $x^a \div x^b = x^{a-b}$

C. $\boxed{10}\,x^7 \div \boxed{2}\,x^4 = \boxed{5}\,x^3$

- Divide the coefficients
- Divide the letters by subtracting exponents

- **Whenever you divide you subtract factors (= why you subtract exponents here)**

- **When dividing you are removing factors**

NOTE: YOU CAN ONLY ADD OR SUBTRACT EXPONENTS IF THE BASES ARE THE SAME!

Divide the following (remember, when you DIVIDE you are REMOVING factors)

1. $a^{10} \div a^2 = a^{\square}$

WITH COEFFICIENTS JUST DIVIDE NORMALLY: 12 ÷ 3!

2. $12a^{10} \div 3a^2 = \square\,\square^{\square}$

3. $\dfrac{y^7}{y^5} = \square^{\square}$

REMEMBER! DIVIDE BY THE DENOMINATOR!

4. $x^8 y^6 \div x^4 y^5 = \square^{\square}\,\square^{\square}$

5. $\dfrac{r^9 s^7}{r^4 s^5} = \square$

6. $\dfrac{15m^6}{5m^2} = \square$

7. $18s^{12}t^{15} \div 3s^9 t^8 = \square$

8. $2a^3 \cdot 6a^5 \div 4a^2 = \square$

9. $\dfrac{24y^9}{3y^4} \cdot 2y^2 = \square$

10. $\dfrac{35a^{12}}{7a^{10}} \cdot 4a^3 = \square$

Exponent Practice

1. $r \cdot r = r^{\square}$

2. $a^5 \div a^2 = \square^{\square}$

3. $y^2 \cdot y^2 \cdot y^2 = y^{\square}$

4. $(x - 3)(x - 3)(x - 3) = (x - 3)^{\square}$

5. $k^2 \cdot k^{10} \div k^3 = \square^{\square}$

6. $b^5 \div b = b^{\square}$

7. $a^5 \cdot a^5 \cdot a^5 = a^{\square}$

8. $\dfrac{d^7}{d^5} = \square^{\square}$

9. $\dfrac{n^7 \cdot n^3}{n^2} = \square^{\square}$

10. $(4x + y)(4x + y)(4x + y) = \boxed{}^{\square}$

11. $yyyy = \square^{\square}$

12. $\dfrac{yyyyy}{yy} = \square^{\square}$

13. $3x^3 \cdot 3x^3 = \square\,\square^{\square}$

14. $2x^5 \cdot 6x^2 = \square\,\square^{\square}$

15. $15x^9 \div 5x^4 = \square\,\square^{\square}$

16. $2x^4 \cdot 2x^4 \cdot 2x^4 = \square\,\square^{\square}$

17. $3y^5 \cdot 6y^2 \div 2y^3 = \square\,\square^{\square}$

18. $2x^2y^3 \cdot 3x^6y^7 = \square\,\square^{\square}\,\square^{\square}$

19. $18a^8b^{10} \div 9a^3b^2 = \square\,\square^{\square}\,\square^{\square}$

Equation Practice

GET RID OF THIS FIRST!

1.   6x - 5 = 19

+ ☐   + ☐

☐ = ☐

REWRITE THE NEW EQUATION!

÷ ☐   ÷ ☐

| x   = |

4.   5b - 3 = 42

+ ☐   + ☐

☐ = ☐

÷ ☐   ÷ ☐

| b   = |

2.   4y + 3 = 35

- ☐   - ☐

☐ = ☐

÷ ☐   ÷ ☐

| y   = |

5.   9x + 5 = -58

- ☐   - ☐

☐ = ☐

÷ ☐   ÷ ☐

| x   = |

3.   7a - 8 = -29

+ ☐   + ☐

☐ = ☐

÷ ☐   ÷ ☐

| a   = |

6.   8c + 4 = 100

- ☐   - ☐

☐ = ☐

÷ ☐   ÷ ☐

| c   = |

© Peter Wise, 2014

35

Equation Practice

Combine like terms and solve for the variable

1. $2x = 15 + 3$

$2x = \boxed{}$

$\div \boxed{} \div \boxed{}$

$x = \boxed{}$

2. $7 + 8 = 5x$

$\boxed{} = 5x$

$\div \boxed{} \div \boxed{}$

$x = \boxed{}$

3. $6x = 38 + 4$

$\boxed{} = \boxed{}$

$\div \boxed{} \div \boxed{}$

$x = \boxed{}$

4. $9x = 22 + 5$

$\boxed{ = }$

$\div \boxed{} \div \boxed{}$

$x = \boxed{}$

5. $27 + 3 = 6x$

$\boxed{ = }$

$\div \boxed{} \div \boxed{}$

DIVIDE BY THE COEFFICIENT OF X!

$\boxed{ = x}$

6. $7x = 30 + 26$

$\boxed{ = }$

$\div \boxed{} \div \boxed{}$

$x = \boxed{}$

7. $11x = 92 - 4$

$\boxed{ = }$

$\div \boxed{} \div \boxed{}$

$x = \boxed{}$

8. $39 + 9 = 8x$

$\boxed{ = }$

$\div \boxed{} \div \boxed{}$

$\boxed{ = x}$

9. $9x = 52 + 11$

$\boxed{ = }$

$\div \boxed{} \div \boxed{}$

$x = \boxed{}$

10. $12x = 31 - 7$

$\boxed{ = }$

$\div \boxed{} \div \boxed{}$

$x = \boxed{}$

11. $9x = 27 + 9$

$\boxed{ = }$

$\div \boxed{} \div \boxed{}$

$x = \boxed{}$

12. $25 - 8 = 17x$

$\boxed{ = }$

$\div \boxed{} \div \boxed{}$

$\boxed{ = x}$

Example

A.

(+ 4x)   (+ 4x)

$$2x = 30 - 4x$$

WHEN YOU ADD 4X TO BOTH SIDES, YOU ADD IT TO 2X!

change sides, change signs

$$\frac{6x}{6} = \frac{30}{6}$$

$$x = 5$$

YOU COULD SUBTRACT 2X FROM BOTH SIDES...BUT IT IS EASIER TO SUBTRACT 4X FROM BOTH SIDES BECAUSE YOU'LL GET A POSITIVE VALUE FOR THE VARIABLE!

Get the letter(s) on one side

Get the number on the other side

Tip: It is usually easiest if you get the bigger (positive) value of the letter on one side and the number on the other side.

IT DOESN'T MATTER WHICH SIDE THE YOU PUT THE LETTER OR NUMBER ON

1. $-2x + 8x = 18$

4. $x + 10 = -2 - 2x$

2. $5x = 21 + 2x$

5. $-3x = -7x + 8$

3. $4x = -10 - 4x - 6$

6. $36 - 2x = x$

Equations: Combining Variables

Combine like terms and solve for the variable

1. $2x + 5x = 14$

$$\boxed{} = 14$$

$(\div)\ \boxed{}\quad (\div)\ \boxed{}$

ONE WAY TO SHOW DIVISION IS BY PUTTING THE DIVISOR UNDER A FRACTION BAR!

$$x = \boxed{}$$

5. $8x - 3x = -55 + 10$

$$\boxed{} = \boxed{}$$
$$\boxed{} \qquad \boxed{}$$

$$x = \boxed{}$$

2. $-8 - 8 = 12x - 4x$

$$\boxed{} = \boxed{}$$
$$\boxed{} \qquad \boxed{}$$

$$x = \boxed{}$$

6. $17x - 15x = 12 + 14$

$$\boxed{} = \boxed{}$$
$$\boxed{} \qquad \boxed{}$$

$$x = \boxed{}$$

3. $18x - 15x = 17 + 16$

$$\boxed{} = \boxed{}$$
$$\boxed{} \qquad \boxed{}$$

$$x = \boxed{}$$

7. $14x - 8x = 22 + 26$

$$\boxed{} = \boxed{}$$
$$\boxed{} \qquad \boxed{}$$

$$x = \boxed{}$$

4. $-30 + 6 = 15x - 7x$

$$\boxed{} = \boxed{}$$
$$\boxed{} \qquad \boxed{}$$

$$x = \boxed{}$$

8. $13x - 4x = 70 - 7$

$$\boxed{} = \boxed{}$$
$$\boxed{} \qquad \boxed{}$$

$$x = \boxed{}$$

Vertical Angle Equations

Example

A.

SINCE VERTICAL ANGLES ARE EQUAL, IT'S JUST LIKE BOTH ANGLES ARE TWO SIDES OF AN EQUATION!

20°

2x

$$\frac{2x}{2} = \frac{20°}{2}$$

$$x = 10°$$

Keep your eyes out for hidden equations like these

Since vertical angles are equal, find the measure of each angle

1.

16°

2x

☐ = ☐ °   x = ☐ °

2.

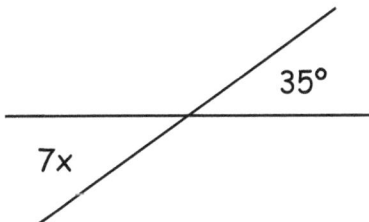

35°

7x

☐ = ☐ °   x = ☐ °

3.

90°

2x

☐ = ☐ °   x = ☐ °

4.

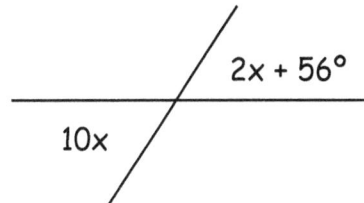

2x + 56°

10x

← Set up your equation

← Subtract from both sides

← Divide from both sides

x = ☐ °

5.

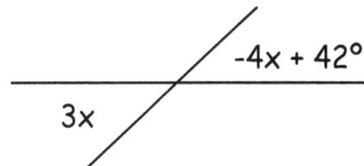

-4x + 42°

3x

x = ☐ °

Equations with Complementary Angles

Complementary Angles Add up to 90°

Examples

A.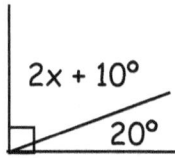
$2x + 10°$
$20°$

Add the values of both angles
and set them equal to 90°

$(2x + 10°) + 20° = 90°$ ← Add the 10° and the 20°

$2x + 30° = 90°$ ← Subtract 30° from both sides

$2x = 60°$ ← Divide both sides by 2

$x = 30°$

B.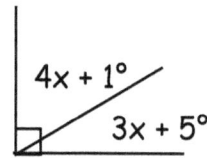
$4x + 1°$
$3x + 5°$

Add the values of both angles
and set them equal to 90°

$(4x + 1°) + (3x + 5°) = 90°$ ← Add the degrees and
the variables

$7x + 6° = 90°$ ← Subtract 6° from both sides

$7x = 84°$ ← Divide both sides by 7

$x = 12°$

Add both angles and set them equal to 90° to find the measure of x

1.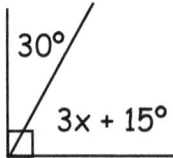
$30°$
$3x + 15°$

← Set up your equation.
Both angles together equal 90°

← Add the degrees on the left
side of the equation

← Eliminate the degrees on the
left side by subtracting

← Divide both sides by the coefficient
(number in front of the variable)

2.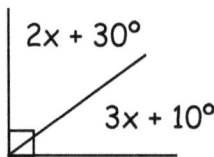
$2x + 30°$
$3x + 10°$

← Set up your equation.
Both angles together equal 90°

← On the left side of the equation add both
pairs of like terms

← Eliminate the degrees on the
left side by subtracting

← Divide both sides by the coefficient
(number in front of the variable)

Now plug in the value of x to
find the measure of each angle:

3.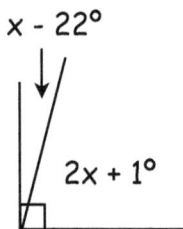
$x - 22°$
$2x + 1°$

← Set up your equation.
Both angles together equal 90°

← On the left side of the equation add both
pairs of like terms

← Eliminate the degrees on the
left side by adding

← Divide both sides by the coefficient
(number in front of the variable)

Now plug in the value of x to
find the measure of each angle:

40

Equations with Supplementary Angles

Supplementary Angles Add up to 180°

A.

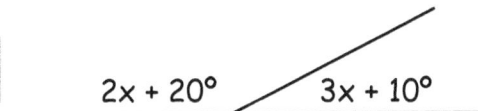

2x + 20°   3x + 10°

Angles: 70° and 20°

Add the values of both angles and set them equal to 180°

$(2x + 20°) + (3x + 10°) = 180°$

$5x + 30° = 180°$ ← Combine the like terms

$5x = 150°$ ← Subtract 30° from both sides

$x = 30°$ ← Divide both sides by 5

Add both angles and set them equal to 180° to find the measure of x

1.

9x - 3°   3x + 15°

← Set up your equation.
Both angles together equal 180°

← Add the like terms on the left
side of the equation

← Eliminate the degrees on the
left side by subtracting

← Divide both sides by the coefficient
(number in front of the variable)

2.

7x + 47°   23x + 43°

← Set up your equation.
Both angles together equal 180°

← Add the like terms on the left
side of the equation

← Eliminate the degrees on the
right side by subtracting

← Divide both sides by the coefficient
(number in front of the variable)

3.

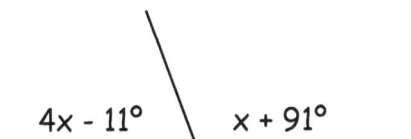

4x - 11°   x + 91°

Now plug in the value of x to
find the measure of each angle:

← Set up your equation.
Both angles together equal 180°

← Add the like terms on the left
side of the equation

← Eliminate the degrees on the
left side by subtracting

← Divide both sides by the coefficient
(number in front of the variable)

Algebra With Perimeters

1.

x (top)
x (left) x (right)
x (bottom)

REMEMBER THAT A LETTER BY ITSELF HAS AN INVISIBLE 1 IN FRONT OF IT!

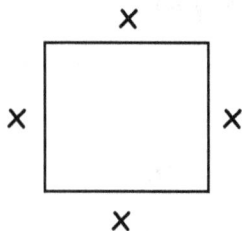

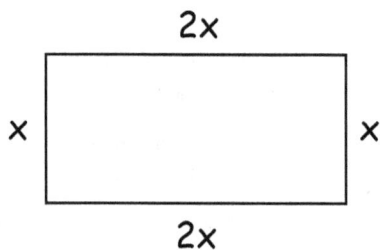

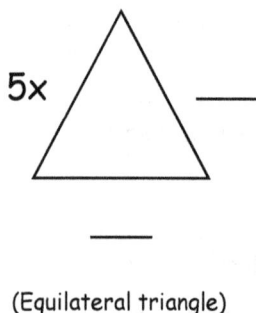

What is the perimeter of this shape, in terms of x? (count around, add up all the variables)

☐ x

If the measurement of the perimeter is 20 inches, what is the value of x?

☐ x = ☐ in   answer: ☐ in

2.

2x (top)
x (left) x (right)
2x (bottom)

What is the perimeter of this shape, in terms of x? (count around, add up all the variables)

☐ x

If the measurement of the perimeter is 18 mm, what is the value of x?

☐ x = ☐ mm   answer: ☐ mm

3.

5x

(Equilateral triangle)

What is the perimeter of this shape, in terms of x? (count around, add up all the variables)

☐ x

If the measurement of the perimeter is 30 cm, what is the value of x?

☐ x = ☐ cm   answer: ☐ cm

Equation Practice

Example

A. $4x + 5 = 25$

Subtract from both sides	-5 -5
Copy the new equation	$4x = 20$
Undo the multiplication (divide both sides)	4 4
Write the value of x	$x = 5$

3. $7x + 8 = 29$

Copy the new equation	
	☐ ☐
	$x =$

1. $6x - 4 = 20$

Add to both sides	
Copy the new equation	
divide both sides	☐ ☐
Write the value of x	$x =$

4. $3x - 3 = 24$

Copy the new equation	
	☐ ☐
	$x =$

2. $6x - 7 = 5$

Copy the new equation	
	☐ ☐
	$x =$

5. $52 = 4 + 8x$

Copy the new equation	
	☐ ☐
	$x =$

Equation Practice

1. $9x + 6 = 78$

Subtract from both sides

$\boxed{}$

Copy the new equation

$\boxed{ = }$

divide both sides

$\boxed{} \quad \boxed{}$

Write the value of x

$\boxed{x = }$

2. $30 = 8y - 10$

$\boxed{}$

Copy the new equation

$\boxed{ = }$

$\boxed{} \quad \boxed{}$

$\boxed{y = }$

3. $7s + 5 = 40$

$\boxed{}$

Copy the new equation

$\boxed{ = }$

$\boxed{} \quad \boxed{}$

$\boxed{s = }$

4. $8g - 13 = 43$

$\boxed{}$

$\boxed{ = }$

$\boxed{} \quad \boxed{}$

$\boxed{g = }$

5. $4x + 7 = 55$

$\boxed{}$

$\boxed{ = }$

$\boxed{} \quad \boxed{}$

$\boxed{x = }$

6. $50 = 6a + 8$

$\boxed{}$

$\boxed{ = }$

$\boxed{} \quad \boxed{}$

$\boxed{a = }$

Equation Practice

Solve the following equations (do the algebra steps!)

1. $94 = 10 + 7s$

Subtract from both sides

Copy the new equation

divide both sides

Write the value of s $s =$

2. $26 = 14 + 4x$

$x =$

3. $12c - 15 = 33$

$c =$

4. $7y + 17 = 80$

$y =$

5. $4n - 20 = 0$

$n =$

6. $73 = 11x + 7$

$x =$

Adding & Subtracting Numerator Variables

Example

A. $\dfrac{3x}{y} + \dfrac{2x}{y} = \dfrac{5x}{y}$ ← Add or subtract numerators

← Copy denominators

Denominators must all match
(no matter what they are)

Add/subtract the following fractions

1. $\dfrac{17n}{r} + \dfrac{3n}{r} = \dfrac{\boxed{}}{r}$

2. $\dfrac{6a}{b} - \dfrac{4a}{b} = \dfrac{\boxed{}}{\boxed{}}$

3. $\dfrac{16y}{m} + \dfrac{8y}{m} - \dfrac{2y}{m} = \dfrac{\boxed{}}{\boxed{}}$

4. $\dfrac{20a}{c-d} + \dfrac{15a}{c-d} - \dfrac{6a}{c-d} = \dfrac{\boxed{}}{c-d}$

5. $\dfrac{38m}{y} - \dfrac{5m}{y} = \dfrac{\boxed{}}{\boxed{}}$

6. $\dfrac{20a}{x^2-y^2} + \dfrac{8a}{x^2-y^2} = \dfrac{\boxed{}}{\boxed{}}$

46

Variables Divided by Numbers

Examples

A. $\dfrac{x}{3} = 2$

$(3)\dfrac{x}{3} = 2(3)$

THE WAY TO UNDO ÷ 3 IS TO MULTIPLY BY 3!

$x = 6$

B. $\dfrac{x}{5} + 2 = 5$

$-\boxed{2} \quad -\boxed{2}$

$\dfrac{x}{5} = 3$

$(5)\dfrac{x}{5} = 3(5)$

$\boxed{x = 15}$

Solve for x by undoing the division

1. $\dfrac{x}{4} = 2$

$(\quad)\dfrac{x}{4} = 2(\quad)$

MULTIPLYING BY THE SAME NUMBER CANCELS OUT THE DENOMINATOR!

$\boxed{x = \qquad}$

2. $\dfrac{x}{10} = 7$

$(\quad)\dfrac{x}{10} = 7(\quad)$

$\boxed{x = \qquad}$

3. $5 = \dfrac{x}{9}$

$(\quad)5 = \dfrac{x}{9}(\quad)$

$\boxed{\qquad = x}$

4. $\dfrac{x}{7} = 1 + 3$

$(\quad)\dfrac{x}{7} = \boxed{}(\quad)$

$\boxed{x = \qquad}$

5. $6 = \dfrac{x}{2}$

$(\quad)6 = \dfrac{x}{2}(\quad)$

$\boxed{\qquad = x}$

6. $\dfrac{y}{8} = 3$

$(\quad)\dfrac{y}{8} = 3(\quad)$

$\boxed{y = \qquad}$

Examples

THIS FRACTION WILL DISAPPEAR WHEN YOU MULTIPLY IT BY ITS RECIPROCAL!

A. $\frac{2}{3}x = 6$

$\left(\frac{3}{2}\right)$ Multiply both sides of the equation by the reciprocal $\left(\frac{3}{2}\right)$

Rule

Fractions disappear (turn to 1) when you multiply them by their reciprocals

$$\left(\frac{3}{2}\right)\frac{2}{3}x = 6\left(\frac{3}{2}\right)$$

×3

÷2

$$x = 9$$

answer

THESE FRACTIONS CANCEL EACH OTHER OUT!

When you multiply a fraction by its reciprocal you always get a new fraction equal to one

---

Solve the following equations by multiplying both sides by the fraction reciprocals

1. $\frac{2}{5}x = 8$

DIVIDE BY THE BOTTOM AND MULTIPLY BY THE TOP!

$$(\text{—})\frac{2}{5}x = 8(\text{—})$$

THE RECIPROCAL (FLIP) OF 2/5 GOES HERE!

$x =$ ☐

3. $\frac{5}{9}x = 15$

$$(\text{—})\frac{5}{9}x = 15\ (\text{—})$$

$x =$ ☐

2. $\frac{3}{4}x = 6$

$$(\text{—})\frac{3}{4}x = 6\ (\text{—})$$

$x =$ ☐

4. $\frac{4}{3}x = 20$

$$(\text{—})\frac{4}{3}x = 20\ (\text{—})$$

$x =$ ☐

Solve the following equations by multiplying both sides by the fraction reciprocals

1. $\frac{7}{3}x = 21$

$(—)\frac{7}{3}x = 21(—)$

$x = \boxed{}$

2. $\frac{5}{6}x = 30$

$(—)\frac{5}{6}x = 30 \ (—)$

$x = \boxed{}$

3. $\frac{3}{8}x = 12$

$(—)\frac{3}{8}x = 12(—)$

$x = \boxed{}$

4. $\frac{5}{4}x = 20$

$(—)\frac{5}{4}x = 20(—)$

$x = \boxed{}$

5. $\frac{8}{3}x = 16$

$(—)\frac{8}{3}x = 16(—)$

$x = \boxed{}$

6. $\frac{6}{7}x = 18$

$(—)\frac{6}{7}x = 18(—)$

$x = \boxed{}$

7. $3\frac{1}{3}x = 40$

$(—)\dfrac{\boxed{}}{\boxed{}}x = 40 \ (—)$

$x = \boxed{}$

8. $3\frac{3}{4}x = 30$

$(—)\dfrac{\boxed{}}{\boxed{}}x = 30 \ (—)$

$x = \boxed{}$

Fraction Coefficient Word Problems

Example

A. 2/3 of a number is 12. What is the number?

$$\frac{2}{3}x = 12$$

$\left(\frac{3}{2}\right)$ Multiply both sides of the equation by the reciprocal $\left(\frac{3}{2}\right)$

× 3

÷ 2

x = [18]

Write the following word problems as equations with fraction coefficients, then solve

1. 3/4 of a number is 9. What is the number?

$\dfrac{\square}{\square}\,\square = \square$ x = $\square$

4. 7/9 of the weight is 14 lbs. What is the total weight?

$\dfrac{\square}{\square}\,\square = \square$ x = $\square$ lbs

2. 5/7 of a number is 15. What is the number?

$\dfrac{\square}{\square}\,\square = \square$ x = $\square$

5. 5/11 of the money is $250. What is the total amount of money?

x = $ $\square$

3. 4/5 of the distance is 20 miles. How far is the total distance?

$\dfrac{\square}{\square}\,\square = \square$ x = $\square$ mi

6. 3/8 of the water is 300 gallons. What is the total amount?

x = $\square$ gal

50

Equation Practice

GET X ON THE SIDE WHERE IT IS GREATER!

X IS GREATER ON THIS SIDE!

SO GET RID OF X ON THIS SIDE!

A. $10x = 3x + 21$

$- \boxed{3x} \quad - \boxed{3x}$

$\boxed{7x \quad = \quad 21}$

$\div \boxed{7} \quad \div \boxed{7}$

$\boxed{x \quad = \quad 3}$

tip: Get x on one side and the numbers on the other side

B. $3x = -2x + 10$

$+ \boxed{2x} \quad + \boxed{2x}$

$\boxed{5x \quad = \quad 10}$

$\div \boxed{5} \quad \div \boxed{5}$

$\boxed{x \quad = \quad 2}$

Get x on the side where it's greater, then solve

1. $10x = 2x + 32$

$- \boxed{} \quad - \boxed{}$

$\boxed{\quad = \quad}$

$\div \boxed{} \quad \div \boxed{}$

$\boxed{x \quad = \quad}$

2. $36x = -4x + 80$

$+ \boxed{} \quad + \boxed{}$

$\boxed{\quad = \quad}$

$\div \boxed{} \quad \div \boxed{}$

$\boxed{x \quad = \quad}$

3. $14x = 12x + 24$

$- \boxed{} \quad - \boxed{}$

$\boxed{\quad = \quad}$

$\div \boxed{} \quad \div \boxed{}$

$\boxed{x \quad = \quad}$

4. $7x = -2x + 27$

$+ \boxed{} \quad + \boxed{}$

$\boxed{\quad = \quad}$

$\div \boxed{} \quad \div \boxed{}$

$\boxed{x \quad = \quad}$

5. $3x = -3x + (30 + 6)$

$+ \boxed{} \quad + \boxed{}$

$\boxed{\quad = \quad}$

$\div \boxed{} \quad \div \boxed{}$

$\boxed{x \quad = \quad}$

6. $17x = 10x + 35$

$- \boxed{} \quad - \boxed{}$

$\boxed{\quad = \quad}$

$\div \boxed{} \quad \div \boxed{}$

$\boxed{x \quad = \quad}$

Equation Practice

1. $4x = 30 + 2x$

   $x = \boxed{}$

2. $-7x = 36 - 11x$

   $x = \boxed{}$

3. $2x - 8 = 40 - 4x$

   $x = \boxed{}$

4. $4x - 70 = -3x + 7$

   $x = \boxed{}$

5. $-2x - 3 = -11x + 60$

   $x = \boxed{}$

6. $10x - 6x - 70 = -8x + 2$

   $x = \boxed{}$

7. $4x + 13 = -5 - 2x$

   $x = \boxed{}$

8. $-2x - 4 = -10x + 60$

   $x = \boxed{}$

9. $-3x + 30 = -10x - 5$

   $x = \boxed{}$

10. $5x - 40 = 8 - 7x$

    $x = \boxed{}$

Intro to Proportions

Example

A. $\dfrac{2}{9} = \dfrac{x}{12}$

#1 Cross multiply $\dfrac{2}{9} \bowtie \dfrac{x}{12}$ $9x = 24$

#2 Divide both sides by the number in front of the letter (the "coefficient") $\rightarrow \dfrac{9x}{9} \rightarrow \dfrac{24}{9}$

#3 Simplify $x = \dfrac{24}{9} = 2\dfrac{6}{9}$

$= \boxed{2\dfrac{2}{3}}$

Solve the following proportions; leave your answers as fractions or mixed numbers

1. $\dfrac{2}{7} = \dfrac{x}{3}$

#1 Cross multiply

#2 Divide both sides by the number in front of the letter (the "coefficient")

$x =$

2. $\dfrac{4}{11} = \dfrac{x}{12}$

#1 Cross multiply

#2 Divide both sides

$x =$

3. $\dfrac{3}{7} = \dfrac{x}{10}$

#1

#2

$x =$

4. $\dfrac{3}{5} = \dfrac{4}{x}$

$x =$

5. $\dfrac{x}{5} = \dfrac{2}{3}$

$x =$

53

Proportion Practice

Solve the following proportions; leave your answers as fractions or mixed numbers

1. $\dfrac{2}{5} = \dfrac{x}{8}$

#1 Cross multiply

#2 Divide both sides by the number in front of the letter (the "coefficient")

x =

2. $\dfrac{3}{4} = \dfrac{x}{7}$

#1 Cross multiply

#2 Divide both sides

x =

3. $\dfrac{2}{x} = \dfrac{7}{6}$

#1

#2

x =

4. $\dfrac{x}{5} = \dfrac{7}{2}$

#1

#2

x =

5. $\dfrac{7}{9} = \dfrac{5}{x}$

x =

6. $\dfrac{4}{9} = \dfrac{x}{5}$

x =

7. $\dfrac{2}{5} = \dfrac{x}{12}$

x =

8. $\dfrac{10}{x} = \dfrac{8}{7}$

x =

SIMPLIFY! x =

Proportion Practice

1. $\dfrac{x}{4} = \dfrac{3}{7}$

x = ☐

2. $\dfrac{n}{3} = \dfrac{5}{8}$

x = ☐

3. $\dfrac{2}{x} = \dfrac{3}{4}$

x = ☐

4. $\dfrac{3}{5} = \dfrac{x}{6}$

x = ☐

5. $\dfrac{8}{3} = \dfrac{x}{4}$

x = ☐

6. $\dfrac{x}{7} = \dfrac{4}{5}$

x = ☐

7. $\dfrac{6}{5} = \dfrac{x}{9}$

x = ☐

8. $\dfrac{x}{100} = \dfrac{3}{4}$

x = ☐

9. $\dfrac{r}{6} = \dfrac{7}{8}$

x = ☐

SIMPLIFY!

10. $\dfrac{x}{8} = \dfrac{6}{7}$

x = ☐

Intro to the Distributive Property

Example

Double burgers and double fries

×2   ×2

2(burgers + fries) = 2 burgers + 2 fries

Copy the + or - signs inside the parentheses ↑

DISTRIBUTING NUMBER

MULTIPLIES EVERYTHING INSIDE THE PARENTHESES!

the burgers are doubled and the fries are doubled

Multiply everything inside the parentheses by the outside distributing number

1. 4(sandwich + drink)

   ☐ _____ + ☐ _____

REMEMBER THE "BURGER-FRIES" TRICK!

2. 5(book + cover)

   ☐ _____ + ☐ _____

3. 2(left shoe + right shoe)

   _____ + _____

4. 3(salad + sandwiches + dessert)

   _____ + _____ + _____

56

Intro to the Distributive Property

1. (desk + lamp)2

   THE DISTRIBUTING NUMBER WORKS THE SAME AT THE END AS IT DOES AT THE FRONT!

   [ ] _____ + [ ] _____

2. 3(dollar - penny)

   [ ] _____ - [ ] _____ = $ _____

   Calculate the amount

3. 4(dozen - 2)

   _____ - _____ = _____

4. 5(dime + nickel + penny)

   _____ ¢ + _____ ¢ + _____ ¢ = _____ ¢

   Calculate amounts in cents

5. (quarter + dime + nickel)3

   $ _____ + $ _____ + $ _____ = $ _____

   Calculate amounts in dollars

© Peter Wise, 2014

57

Practice with the Distributive Property

Multiply everything inside the parentheses by the outside distributing number

Calculate Money Amounts

1.   4($2 - 1¢)

_____ - _____ = [   ]

> DID YOU REMEMBER TO MULTIPLY **BOTH** INSIDE NUMBERS BY 4?

2.   6(dime + nickel)

_____ + _____ = [   ]

3.   (quarter + dime + penny)2

_____ + _____ + _____

= [   ]

4.   3($10 bill + $5 bill + $1 bill)

_____ + _____ + _____

= [   ]

Calculate Number Amounts

5.   3(20 + 5)

_____ + _____ = [   ]

6.   (12 + 5)4

_____ + _____ = [   ]

7.   5(6 + 10)

_____ + _____ = [   ]

8.   7(8 + 9)

_____ + _____ = [   ]

Example

A. $2(10 + 7)$

x2 x2

Copy the + or - signs inside the parentheses

$2(10 + 7)$ •2

| 2 • | 10 | + | 2 • | 7 |

20 + 14 = 34

Multiply everything inside the parentheses by the outside distributing number

1. $3(10 + 4)$

$3(10 + 4)$ ×3

☐ ____ + ☐ ____

____ + ____ = ____

2. $2(100 + 1)$

☐ ____ + ☐ ____

____ + ____ = ____

3. $2(100 - 1)$

THIS IS A GREAT WAY TO MULTIPLY 2 X 99!

____ - ____ = ☐

4. $(10 + 4)7$

____ + ____

= ☐

5. $(12 + 3)5$

____ + ____

= ☐

6. 3(1 dollar - 2 cents)

____ - ____ = ☐

The Distributive Property with Variables

the 2 multiplies everything in the parentheses

the a multiplies everything in the parentheses

A. $2(a + b)$

$= \boxed{2a + 2b}$

B. $a(c + d)$

$= \boxed{ac + ad}$

WHEN THEY TOUCH, THEY TIMES!

Expand, using the Distributive Property

Note that you are just rewriting these expressions, not finding an "answer"

1. $3(b + c)$

$3\ \square\ +\ 3\ \square$

2. $7(x + y + z)$

$\square\ x\ +\ \square\ y\ +\ \square\ z$

3. $.2(a + b)$

$\boxed{}\ +\ \boxed{}$

4. $\dfrac{1}{2}(m + r)$

$\boxed{}\ +\ \boxed{}$

5. $5c(c + d)$

$\boxed{}\ +\ \boxed{}$

WATCH CAREFULLY FOR EXPONENTS!

6. $(r + s)r^2$

THE DISTRIBUTING NUMBER CAN BE IN FRONT OR IN BACK— IT WORKS THE SAME EITHER WAY!

$\boxed{}\ +\ \boxed{}$

7. $a^3(a^2 + ab - 2)$

$\boxed{}$

8. $x^2y^3(x^5 - y^3)$

$\boxed{}$

The Distributive Property with Variables

Examples

$a \cdot a = a^2$ $a \cdot b = ab$

WHEN THEY TOUCH, THEY TIMES!

$a \cdot 2 = 2a$

IT IS COMMON TO PUT THE LETTER AFTER THE NUMBER!

A. $a\,(a + b)$

$= \boxed{a^2 + ab}$

B. $x\,(2 + y)$

$= \boxed{2x + xy}$

Expand, using the Distributive Property

Note that you are just rewriting these expressions, not finding an "answer"

1. $m(m + n)$

$\boxed{} + \boxed{}$

5. $(x + y)x$

$\boxed{}\ \boxed{}\ \boxed{}$

2. $n(n - p)$

$\boxed{} - \boxed{}$

6. $(m - n)n$

$\boxed{}\ \boxed{}\ \boxed{}$

3. $r(2r + s)$

$2\ \boxed{}\ \boxed{}\ \boxed{}$

PUT THE CORRECT SIGN HERE!

7. $s(r + s - t)$

$\boxed{}\ \boxed{}\ \boxed{}\ \boxed{}\ \boxed{}$

4. $y(y + 3)$

$\boxed{}\ \boxed{}\ \boxed{}$

8. $x(4 - x + y)$

$\boxed{}\ \boxed{}\ \boxed{}\ \boxed{}\ \boxed{}$

Distributing a Negative Sign

Examples

A NEGATIVE DISTRIBUTING NUMBER OR LETTER *REVERSES ALL THE SIGNS INSIDE PARENTHESES!*

A. $-x(x + y - z)$

$$-x^2 - xy + xz$$

A NEGATIVE SIGN BY ITSELF OUTSIDE THE PARENTHESES IS REALLY (-1)!

B. $-(x + y - z)$

$$-x - y + z$$

IN THIS CASE, THE SIGNS JUST SWITCH, BUT EVERYTHING ELSE STAYS THE SAME!

C. $20 - (2 + 3 + 4)$

20 | $-2 - 3 - 4$

$= 20 - 9 = 11$

Anything negative outside the parentheses is a sign-switcher

Expand, using the Distributive Property

IT SWITCHES THE SIGNS OF ALL THE NUMBERS INSIDE THE PARENTHESES!

1. $-(a - b + c - d)$

THE NEGATIVE NUMBER OUTSIDE IS A SIGN SWITCHER!

2. $-x(x - y - z)$

3. $-2a(2a + 3b - 4c)$

4. $-10x^2(2x^3 - 3y^2)$

5. $-6m(-3m + 2m^2)$

6. $-4g^2h^3(5h - 3g^5h^2)$

7. $-7r^4s^5(2r^2s^6 - 3r^5s^8)$

8. $-5m(5m - 7 - 6mn^3)$

62

Practice with the Distributive Property

Expand, using the Distributive Property

1. $x(x + y)$

+

2. $x(x^2 + y^2)$

+

3. $5c(c - d)$

-

Write the addition or subtraction signs

↓

4. $(f + g)7f^3$

REMINDER!
A NEGATIVE DISTRIBUTING NUMBER ACTS AS A SIGN SWITCHER FOR THE VALUES INSIDE THE PARENTHESES!

5. $-r(3 + s - 10r)$

A NEGATIVE SIGN IS THE SAME AS (-1) MULTIPLYING A NUMBER OR LETTER!

6. $4(x + y)$

7. $2(3n - m)$

8. $b^2(a + b)$

9. $7x(x + y)$

10. $-2a(4 - 3a)$

11. $5r^2(12 + r^5)$

12. $(9m + 7)4n^3$

63

Practice with the Distributive Property

Expand, using the Distributive Property

1. $6s^3(5 - 7s)$

2. $3y^4(9 - 6y^4)$

3. $3c^3(5 - 7s)$

4. $-10c^6(-5d^2 + 4c)$

5. $9g^5(3g^7 - 2h^4)$

6. $(3x^4 - 5y^3)7x^2$

> HINT:
> DISTRIBUTING NUMBERS HAVE THE SAME EFFECT WHETHER THEY'RE IN BACK OR FRONT OF PARENTHESES!

7. $8(7x - 3y)$

8. $(4a + 2b - 3c)9$

9. $12m^5(4 - 7m^2)$

10. $7r(7r + 4r^2)$

11. $-4x^7(3x^2 - 8y^5)$

12. $12n^4(2m^3 - 9n^2)$

Distributive Property & Combining Like Terms

Expand, using the Distributive Property, then combine like terms

1. $5(3a + b) + 2(a + 4b)$

    #1 Expand, using the Distributive Property:

+		+	+

    $5(3a + b)$ $2(a + 4b)$

    #2 Combine like terms: [ ]a + [ ]b

2. $3(x + y) + 4(x + y)$

    #1 Expand, using the Distributive Property:

+		+	+

    #2 Combine like terms: [ ]x + [ ]y

3. $7(2m + 3r) + 3(2m + 10r)$

    #1 Expand, using the Distributive Property: [+] + [+]

    #2 Combine like terms: [ ] + [ ]

4. $3(2x - 5y) + 2(9y - 12x)$

    #1 Expand, using the Distributive Property: [-] + [-]

    #2 Combine like terms: [ ] + [ ]

The Distributive Property with Equations

A.

$2(x + 3) = 16$

$2x + 6 = 16$ Expand, using the Distributive Property

$2x = 10$ Subtract 6 from both sides

$x = 5$ Divide both sides by 2

Expand, using the Distributive Property, then solve the equations

1. $2(x + 4) = 14$ x = ☐

    ← Expand, using the Dist. Prop.

    ← Subtract from both sides

    ← Divide both sides

4. $6(x - 2) = 48$ x = ☐

2. $3(x + 2) = 24$ x = ☐

    ← Expand, using the Dist. Prop.

    ← Subtract from both sides

    ← Divide both sides

5. $-2(x + 3) = -14$ x = ☐

3. $2(x - 5) = 6$ x = ☐

6. $-3(x - 4) = -15$ x = ☐

Algebra With Perimeters

Example

A. The perimeter of the rectangle is 36. What is the value of x?

$2(3x + 1) + 2(4x + 3) = 36$
Double length   Double width

$(6x + 2) + (8x + 6) = 36$

$14x + 8 = 36$

$14x = 28 \qquad x = 2$

w = 3x + 1

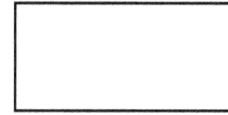

Double this value for the width (or height)

l = 4x + 3

Double this value for the length (base)

PERIMETER MEASURES THE DISTANCE AROUND A 2-DIMENSIONAL SHAPE!

Perimeter of parallelograms = Double length + double width

2L + 2W!

---

1. The perimeter of the rectangle is 46 units. What is the value of a?

You double each dimension for the perimeter of parallelograms

2x + 2

3x + 5

☐ (☐) + ☐ (☐)

☐ + ☐

☐ = 44 units   x = ☐

---

2. The perimeter of the rectangle is 80 cm. What is the value of y?

2y + 3

4y + 7

☐ (☐) + ☐ (☐)

☐ + ☐

☐ = 80 units

x = ☐

67

Factoring: Reverse Distributive

Example

A.

$$10 + 15$$
$$\uparrow \qquad \uparrow$$
$$5(2 + 3)$$

5 TIMES 2 = 10!

5 TIMES 3 = 15!

Factor, using the Distributive Property

1.

$$6 + 9$$
$$\uparrow \qquad \uparrow$$
$$3(\boxed{} + \boxed{})$$

What do you multiply 3 by to get 6?   What do you multiply 3 by to get 9?

5.

$$44 + 55$$
$$\uparrow \qquad \uparrow$$
$$\boxed{}(\boxed{} + \boxed{})$$

WHAT NUMBER DIVIDES EVENLY INTO 44 AND 55?

What do you multiply the distributing number by to get 44?   ...to get 55?

2.

$$21 - 28$$
$$\uparrow \qquad \uparrow$$
$$7(\boxed{} - \boxed{})$$

6.

$$63 - 27$$
$$\uparrow \qquad \uparrow$$
$$9(\boxed{} - \boxed{})$$

3.

$$40 + 60 + 80$$
$$\uparrow \qquad \uparrow \qquad \uparrow$$
$$20(\boxed{} + \boxed{} + \boxed{})$$

7.

$$70 + 80$$
$$\uparrow \qquad \uparrow$$
$$\boxed{}(\boxed{} + \boxed{})$$

4.

$$30 + 54$$
$$\uparrow \qquad \uparrow$$
$$6(\boxed{} + \boxed{})$$

8.

$$24 - 16 + 32$$
$$\uparrow \qquad \uparrow \qquad \uparrow$$
$$8(\boxed{} \ \square \ \boxed{} \ \square \ \boxed{})$$

PUT THE CORRECT SIGNS BETWEEN THE BOXES!

Factoring: Reverse Distributive

Example

A. Can divide out $\boxed{5}$ from both

$$15a^3 + 20a^5$$

Can divide out $\boxed{a^3}$ from both
Look for the LOWEST POWER

$\boxed{5a^3}(3 + 4a^2)$

$5a^3 + 20a^5$

These are the largest factors you can pull out   ...this becomes the distributing number

---

Factor out the largest value from each to rewrite with the Distributive Property

With exponents. . .
Look for the lowest power

---

1. $6a^2 + 8a^3$

[ ] ← largest number that divides into both

[ ] ← largest power of a letter that divides into both... look for the LOWEST POWER!

[ ] $(3 + 4a)$

THE NUMBER AND LETTER ABOVE FORM THE GCF. PUT BOTH OF THEM HERE FOR THE DISTRIBUTING TERM!

3. $14n^3 + 21n^5$

[ ] ← largest number that divides into both

[ ] ← largest power of a letter that divides into both...

[ ] $(2 + \boxed{}\, n^{\boxed{}})$

2. $12y + 15y^2$

[ ] ← largest number that divides into both

[ ] ← largest power of a letter that divides into both... look for the LOWEST POWER!

[ ] $(4 + 5y)$

4. $30m^7 - 70m^{10}$

[ ] ← largest number that divides into both

[ ] ← largest power of a letter that divides into both...

[ ] $(3 - \boxed{}\, m^{\boxed{}})$

69

Factoring: Reverse Distributive

A.

Can divide out $\boxed{3}$ from both

$$6y^3 + 9y^2$$

Can divide out $\boxed{y^2}$ from both
Look for the LOWEST POWER

$$\boxed{3y^2}(2y + 4)$$

$6y^3 + 9y^2$

These are the largest factors you can pull out

...this becomes the distributing number

Factor out the largest value from each to rewrite with the Distributive Property

1. $18a^4 + 27a^3$

$\boxed{}$ ← largest number that divides into both

$\boxed{}$ ← largest power of a letter that divides into both... look for the LOWEST POWER!

$\boxed{} (\boxed{} + \boxed{})$

3. $12m^{10} + 20m^7$

$\boxed{}$ ← largest number that divides into both

$\boxed{}$ ← largest power of a letter that divides into both... look for the LOWEST POWER!

$\boxed{} (\boxed{} + \boxed{})$

2. $24c^5 + 36c^9$

$\boxed{}$ ← largest number that divides into both

$\boxed{}$ ← largest power of a letter that divides into both... look for the LOWEST POWER!

$\boxed{} (\boxed{} + \boxed{})$

4. $21x^5 + 28x^2 - 14x^3$

$\boxed{}$ ← largest number that divides into both

$\boxed{}$ ← largest power of a letter that divides into both... look for the LOWEST POWER!

$\boxed{} (\boxed{} + \boxed{} - \boxed{})$

Factoring Practice

1. $12r^5 - 8r^2$

[　　] ← largest number that divides into both

[　　] ← largest power of a letter that divides into both... look for the LOWEST POWER!

[　　] ([　　] - [　　])

2. $5c^5 + 15c^6$

[　　] ([　　] + [　　])

↑ largest number AND power of a letter that divide into both

3. $24s^7 + 27s^{10} - 18s^9$

[　　] ([　　] + [　　] - [　　])

↑ largest number AND power of a letter that divide into both

4. $63m^4n^3 + 45m^2n^5$

[　　] ← largest number that divides into both

[　　] ← largest power of m that divides into both... look for the LOWEST POWER!

[　　] ← largest power of n that divides into both... look for the LOWEST POWER!

[　　] ([　　] - [　　])

5. $33a^2b^{11} - 77a^7b^3$

[　　] ([　　] - [　　])

6. $20x^6y^{10} - 28x^4y^{15}$

[　　] ([　　] - [　　])

Challenge Question:

7. $-3a^2 - 6y^2 - 5x^2$

[　] ([　　　　　　])

Hint: Look for the sign-switcher

Letters on One Side, Numbers on the Other

It actually doesn't matter which side you put the letters or numbers on

Simplest way: Look for the side with the greatest value for the letter.

Put the letter(s) on that side and the number(s) on the other side

Solve by getting the letters on one side and the numbers on the other side

Letters this side   Numbers ("constants") this side

1.

$$9x - 10 = 4x + 5$$

x = ☐

The larger value of x is on this side. It's a little simpler to get the letter(s) on this side.

← Subtract 4x from both sides (gets the variable out of the right side)

← Add 10 to both sides (gets rid of the number on the left side)

← Divide both sides by the number multiplying the letter (the "coefficient")

Constants this side   Letters this side

2.

$$-3x + 9 = 4x - 5$$

x = ☐

The larger value of x is on the RIGHT side. It's a little simpler to get the letter(s) on this side.

← Add 3x to both sides (gets the variable out of the left side)

← Add 5 to both sides (gets rid of the negative number on the right side)

← Divide both sides by the number multiplying the letter (the "coefficient")

3.

$$2x + 2 = -2x + 30$$

x = ☐

← Add 2x to both sides (gets the variable out of the left side)

← Subtract 2 from both sides (gets rid of the number on the left side)

← Divide both sides by the coefficient

4.

$$4x - 4 = 7x - 28$$

$x =$ ☐

← Subtract 4x from both sides (gets the variable out of the left side)

← Add 28 to both sides (gets rid of the number on the right side)

← Divide both sides by the number multiplying the letter (the "coefficient")

5.

Letters this side   Constants this side

$$16x - 5 = 2x + 40 + 5x$$

$x =$ ☐

← Combine the x-values on the right side

← Subtract 7x from both sides (gets the variable out of the right side)

← Add 5 to both sides (gets rid of the negative number on the right side)

← Divide both sides by the number multiplying the letter (the "coefficient")

6.

$$2x + 4x + 5x - 4 = 3x + 28$$

$x =$ ☐

← Combine the x-values on the left side

← Subtract 3x from both sides (gets the variable out of the right side)

← Add 4 to both sides (gets rid of the constant on the left side)

← Divide both sides by the number multiplying the letter (the "coefficient")

7.

$$12x - 22 = 3x + 5$$

$x =$ ☐

← Subtract 3x from both sides (gets the variable out of the right side)

← Add 22 to both sides (gives you constants on the right side)

← Divide both sides by the number multiplying the letter (the "coefficient")

73

Multi-Step Equation Practice

Solve by getting the letters on one side and the numbers on the other side

1.

Variable this side   Numbers this side

$4(3x + 2) - 1 = -29$

x = ☐

← Expand, using the Distributive Property

← Combine the like terms on the left side

← Eliminate numbers on the left side

← Divide by the coefficient

2.

Variable this side   Numbers this side

$5x + 2(5 - 3) = -4 + 8$

x = ☐

← Expand, using the Distributive Property

← Combine the like terms on the left side

← Get the numbers all on the right side

← Divide by the coefficient

3.

Numbers this side   Variable this side

$14x - 10 - 2 = 4(4x - 2) + 22$

x = ☐

← Expand, using the Distributive Property

← Combine the constants on both sides

← Get the variable only on the right side
(because the value of x is larger on that side)

← Eliminate the constant on the right side

← Divide by the coefficient

Equations: Variables on Both Sides

1. $4x = 32 + 2x$

$x =$ ☐

2. $-7x = 36 - 11x$

$x =$ ☐

3. $2x - 8 = 40 - 4x$

$x =$ ☐

4. $4x - 70 = -3x + 7$

$x =$ ☐

5. $-2x - 3 = -11x + 60$

$x =$ ☐

6. $8x - 7x - 40 = -6x + 2$

$x =$ ☐

7. $4x + 13 = -5 - 2x$

$x =$ ☐

8. $3x - 4 = -2x + 60 - 64$

$x =$ ☐

9. $-3x + 30 = -10x - 5$

$x =$ ☐

10. $5x - 40 = 8 - 7x$

$x =$ ☐

1. If a letter has no number in front if it, the invisible number is really ☐

2. Terms that have matching letters and matching exponents are called this

(Example: $4x^2y^3$ and $5x^2y^3$)

3. Write the equation for the following:

A number doubled and increased by 2 equals the same number tripled and decreased by 7

4.   35°

_____   Give the value for x ☐ °

8x - 13°

5. Use substitution to solve:

$b^2 - 4ac$ = ☐

$a = 2$ $b = 9$ $c = 8$

6. Use substitution to solve:

$(x^2 + 2x)mn$ = ☐

$x = 3$ $m = 2$ $n = 4$

Circle the like terms

7. $14y^3x^4$ $14x^3y^4$ $14x^4y^3$ $3y^3x^4$

Simplify problems 8-12

8. $9x - 5x + 7x =$ ☐

9. $a^2b^5 \cdot a^4b^2c^3 =$ ☐

10. $5a^3b^6 \cdot 7a^2b^5 \cdot b =$ ☐

11. $12x^7y^8 \div 4x^3y^3 =$ ☐

12. $\dfrac{5y}{x} + \dfrac{9y}{x} - \dfrac{2y}{x} =$ ☐

13. Your goal in solving equations is to
_____ the variable

14. $8x + 18 = 74$ $x =$ ☐

Show your steps!

15. $13 + 15 = 22x - 15x$ $x =$ ☐

Show your steps!

16. $\dfrac{x}{9} = 6$ $x =$ ☐

17. $-10x + 2x - 6x + 5x =$ ▢

18. Exponents indicate the number of repeated _____

19. $3x = -9x + 48$ $x =$ ▢

Show your steps!

20. $\frac{3}{7}x = 12$ $x =$ ▢

21. Solve the following proportion. Leave your answer as a mixed number in simplest form $\frac{3}{8} = \frac{x}{12}$ $x =$ ▢

22. $9(a + b - c) =$ ▢

23. $(r^3 - s^4 + t^2)s^2$ ▢

24. $-2b^3(7a^5 - 3b^5 + 4c^2)$ ▢

25. $8(x - 3) = 48$ ▢

26. Factor (reverse Distributive Property)

21 - 35 $\boxed{}\left(\boxed{} - \boxed{}\right)$

27. Factor (reverse Distributive Property)

$15y^2 + 20y^3$ $\boxed{}\left(\boxed{} + \boxed{}\right)$

Constants this side   Letters this side

28. $-3x + 9 = 4x - 5$ x = $\boxed{}$

← Add 3x to both sides
(gets the variable out of the left side)

← Add 5 to both sides (gets rid of the
negative number on the right side)

← Divide both sides by the number
multiplying the letter (the "coefficient")

29. $-2x - 4 = -10x + 60$ x = $\boxed{}$

30. $-6x - 13 = -15x + 50$ x = $\boxed{}$

Algebra Quiz

1. $y + y + y = $ ☐

2. $(y)(y)(y) = $ ☐

3. $15c + 15c - 6c = $ ☐

4. $3x^4 \cdot 7x^5 = $ ☐

5. $5a^3b^6 \cdot 7a^2b^4 = $ ☐

6. $10a^9b^{12} \div 2a^2b^8 = $ ☐

7. Factor $18a^2 - 27a^3$ ☐

(Largest possible factors)

8. $2a^2 \cdot 2a^3 \cdot 3a^4 = $ ☐

9. $2(x - 5) = 6$ $x = $ ☐

Show your steps!

10. $12x^5$ $12x^3$ $4x^3$

Circle the like terms

11. Use substitution to solve:

$4ac = $ ☐ $a = 6$
$c = 2$

12. Use substitution to solve:

$2b^2 = $ ☐ $b = 9$

13. $8x + 12 = 44$ $x = $ ☐

Show your steps!

14. $10x + 5x = 3x + 24$ $x = $ ☐

Show your steps!

15. $\dfrac{5}{9} = \dfrac{x}{3}$ $x = $ ☐

Express your answer
as a mixed number
in simplest form

16. $\dfrac{x}{5} = 7$ $x = $ ☐

Answer Key

for

MathWise Algebra, Book 1

Basic Algebra Terms

STUDY THESE ANSWERS AND THEN DO THE NEXT PAGE

1.   a, x, y

Letters in algebra are called:

variables   because

their values vary

2.   We usually don't use x for multiplication in higher math because

x as a letter can be confused with x for multiplication

3.   There are three ways to show multiplication:

A.   **x, as in 3 x 4**   *(THIS SYMBOL IS USED MOSTLY IN LOWER GRADES!)*

B.   **dot, as in 3 • 4**

C.   **parentheses: 3(4), (3)4, or (3)(4)**

4.   Rule: "When they touch..."   **"they times" (multiply)**

5.   ⑤x = 15

WOW! THAT'S A BIG WORD! AM I IN COLLEGE ALREADY?

When a number is touching (multiplying) a letter it is called a

coefficient

6.   You can write 3 times 4 three different ways using parentheses:

3(4)   **(3)4**

(3)(4)

7.   You use a letter (doesn't matter which) when you see these words in a problem:

"a number"

"some number"

8.   Use an equal sign when you see these words:

"is"   **"the result is"**

© Peter Wise, 2014

1

Basic Algebra Terms

STUDY THESE ANSWERS AND THEN DO THE NEXT PAGE

1.   a, x, y

Letters in algebra are called:

variables   because

their values vary

2.   We usually don't use x for multiplication in higher math because

x as a letter can be confused with x for multiplication

3.   There are three ways to show multiplication:

A.   **x, as in 3 x 4**   *(THIS SYMBOL IS USED MOSTLY IN LOWER GRADES!)*

B.   **dot, as in 3 • 4**

C.   **parentheses: 3(4), (3)4, or (3)(4)**

4.   Rule: "When they touch..."   **"they times" (multiply)**

5.   ⑤x = 15

WOW! THAT'S A BIG WORD! AM I IN COLLEGE ALREADY?

When a number is touching (multiplying) a letter it is called a

coefficient

6.   You can write 3 times 4 three different ways using parentheses:

3(4)   **(3)4**

(3)(4)

7.   You use a letter (doesn't matter which) when you see these words in a problem:

"a number"

"some number"

8.   Use an equal sign when you see these words:

"is"   **"the result is"**

© Peter Wise, 2014

2

Introducing Variables

- Variables are letters, for example: "a" or "x" or "y"
- They stand for numbers ("mystery numbers" that you have to figure out)
- In different problems variables represent different numbers
- The numbers VARY (that's why they are called VARIables)

Examples

You may have seen math problems like...

$7 + \boxed{} = 10$ $\boxed{} = 3$

$7 + ? = 10$ $? = 3$

IN THE SAME WAY YOU CAN HAVE...

$7 + x = 10$ $x = 3$

A BOX, A QUESTION MARK, OR A LETTER ALL HAVE THE SAME IDEA, BUT IN ALGEBRA WE USUALLY USE LETTERS!

Find the number that the variable represents

1. $x + 2 = 8$ $x =$ **6**

2. $3 + y = 7$ $y =$ **4**

3. $5 + c = 12$ $c =$ **7**

4. $a + 6 = 14$ $a =$ **8**

5. $10 - x = 8$ $x =$ **2**

6. $16 - a = 11$ $a =$ **5**

7. $y - 3 = 6$ $y =$ **9**

8. $7 + b = 15$ $b =$ **8**

9. $14 + x = 20$ $c =$ **6**

10. $40 - d = 31$ $d =$ **9**

11. $18 - x = 15$ $x =$ **3**

12. $a + a = 16$ $a =$ **8**

© Peter Wise, 2014

3

Introducing Variables: Multiplication

- Variables, like numbers, can be added, subtracted, multiplied, or divided

Find the number that the variable represents (multiplication)

1. $3 \cdot y = 12$ $y =$ **4**

   A DOT IS USED IN ALGEBRA, BECAUSE "x" IS USUALLY USED AS A VARIABLE!

2. $x \cdot 2 = 14$ $x =$ **7**

3. $4 \cdot n = 44$ $n =$ **11**

4. $9 \cdot r = 27$ $r =$ **3**

5. $a \cdot 2 = 24$ $a =$ **12**

6. $8 \cdot y = 56$ $y =$ **7**

7. $6 \cdot c = 54$ $c =$ **9**

8. $x \cdot 5 = 35$ $x =$ **7**

9. $m \cdot 3 = 24$ $m =$ **8**

10. $a \cdot a = 36$ $a =$ **6**

   SAME LETTERS = SAME NUMBERS

Find the number that the variable represents—WATCH THE SIGNS!

11. $a + 3 = 12$ $a =$ **9**

12. $b \cdot 3 = 33$ $b =$ **11**

13. $4 + r = 10$ $r =$ **6**

14. $e - 2 = 8$ $e =$ **10**

15. $y + 10 = 23$ $y =$ **13**

16. $d \cdot 7 = 42$ $d =$ **6**

© Peter Wise, 2014

Introducing Coefficients

Examples

- Coefficients are numbers right before letters—like 2a
- The rule is "if they touch, they times!"
- Coefficients MULTIPLY the letters next to them

   HERE, 2 IS THE COEFFICIENT AND A IS THE VARIABLE

$2a$ = 2 times a ab = a times b
   $(2 \cdot a)$ $(a \cdot b)$

Use the examples above to solve the following problems

1. $5a =$ **5** $\cdot$ a

2. $3y =$ **3** $\cdot$ y

3. $7x =$ **7** $\cdot$ x

4. $10n =$ **10** $\cdot$ n

5. $14r =$ **14** $\cdot$ r

6. $yz =$ **y** $\cdot$ **z**

LITTLE HARDER...

7. $3 \cdot y =$ **3y**

   NOW WRITE THESE AS THE NUMBER AND TERM TOUCHING!

8. $12 \cdot s =$ **12s**

9. $\frac{1}{2} \cdot a =$ **$\frac{1}{2}$ a**

10. $x \cdot y =$ **xy**

LITTLE HARDER...

11. $a \cdot b \cdot c =$ **abc**

12. $fgh =$ **f** $\cdot$ **g** $\cdot$ **h**

13. $2mn =$ **2** $\cdot$ **m** $\cdot$ **n**

14. $\frac{3}{4} rs =$ **$\frac{3}{4}$** $\cdot$ **r** $\cdot$ **s**

© Peter Wise, 2014

6

Coefficients

Examples

- Remember! Coefficients MULTIPLY the LETTERS next to them
- Multiplication is REPEATED ADDITION

$5 + 5 + 5$ = $3 \cdot 5$

$y + y + y$ = $3 \cdot y$ = $3y$

   YOU ADD THREE Y'S! ... SO IT'S 'THREE TIMES Y!'

Use the examples above to solve the following problems

1. $a + a =$ **2** a

   HOW MANY A'S ARE BEING ADDED?

2. $s + s + s =$ **3** s

3. $4n =$ **n + n + n + n**

4. $x + x + x + x + x =$ **5x**

5. $m + m =$ **2m**
   $+$
   $m + m + m =$ **3m**

   How many m's total (in both lines)? **5m**

6. $4c =$ **c + c + c + c**
   $+$
   $2c =$ **c + c**

   How many c's are being added? **6** c

PUTTING IT ALL TOGETHER...

11. $(n + n + n) + (n + n + n + n) =$

   3 n $+$ **4** n $=$ **7** n

   HOW MANY N'S TOTAL ARE BEING ADDED ?

© Peter Wise, 2014

7

Add the Numbers, Copy the Letters

Examples

A. 2a $+$ 3a $=$ 5a

   THE LETTER STAYS THE SAME!

   RULE: Add the numbers Copy the letters

The numbers touching letters are called **coefficients**

B. $\begin{array}{r} 4a \\ + 3a \\ \hline 7a \end{array}$

   YOU CAN ALSO ADD THESE VERTICALLY!

Add or subtract the following variables

1. $3a + 4a =$ **7** a

2. $5y + 5y + 6y =$ **16** y

3. $12n + 7n =$ **19n**

4. $50r + 20r + 6r =$ **76r**

5. $100x + 100x + 30x + 40x =$ **270x**

6. $600a + 700a + 8a + 40a =$ **1348a**

7. $30y + 4000y + 50y + 2y =$ **4082y**

8. $\begin{array}{r} 5x \\ + 2x \\ \hline 7 \end{array}$ x

9. $\begin{array}{r} 37a \\ + 45a \\ \hline 82a \end{array}$

10. $\begin{array}{r} 389m \\ + 643m \\ \hline 1032m \end{array}$

© Peter Wise, 2014

Subtract the Numbers, Copy the Letters

Examples

THE LETTER STAYS THE SAME!

A. $\boxed{8}a - \boxed{2}a = \boxed{6}a$

RULE:
Subtract the numbers
Copy the letters

The numbers touching letters are called coefficients

B.
$$\begin{array}{r} 9x \\ - 5x \\ \hline 4x \end{array}$$

YOU CAN ALSO SUBTRACT THESE VERTICALLY!

Add or subtract the following variables

1. $20x - 3x = \boxed{17}x$

2. $100y - 10y = \boxed{90}y$

3. $12a - 4a = \boxed{8a}$

4. $17n - 3n = \boxed{14n}$

5. $43z - 4z = \boxed{39z}$

6. $10r + 8r - 3r = \boxed{15r}$

HERE YOU BOTH ADD AND SUBTRACT THE NUMBERS!

7. $20y + 20y - 2y = \boxed{38y}$

8. $15c + 15c - 6c = \boxed{24c}$

9. $28a + 10a - 1a = \boxed{37a}$

10. $14m + 14m - 6m = \boxed{22m}$

11.
$$\begin{array}{r} 35a \\ - 17a \\ \hline \boxed{18}a \end{array}$$

12.
$$\begin{array}{r} 63n \\ - 25n \\ \hline \boxed{38n} \end{array}$$

© Peter Wise, 2014

8

Adding and Subtracting Variables

A. $2x + 3x$

The numbers touching letters are called $\boxed{\text{coefficients}}$

You add or subtract the coefficients, but keep the variables the same.

add just the numbers: 2 + 3 = 5

$2x + 3x = 5x$

$\boxed{x + x}$ $\boxed{x + x + x}$ $\boxed{\begin{array}{c}x + x + x \\ x + x\end{array}}$

$2x + 3x = 5x$

the variable stays the same

Add or subtract the following variables

1. $4x + 3x = \boxed{7x}$

2. $12x - 4x = \boxed{8x}$

3. $5x + 14x = \boxed{19x}$

4. $6a - 2a = \boxed{4a}$

5. $10y + 2y + 4y = \boxed{16y}$

6. $12r - 5r + 20s - 4s =$ $\boxed{7}r + \boxed{16}s$

LETTERS BY THEMSELVES

Letters that are by themselves have an invisible 1 in front of them: $\boxed{y = 1y}$

$y + y = 2y$
$1y + 1y = 2y$

THESE ARE THE SAME!

7. $15x + x = \boxed{16}x$

8. $8a - a + 3b + b = \boxed{7}a + \boxed{4}b$

© Peter Wise, 2014

9

The Invisible ONE

A. $\boxed{1a} + \boxed{1a}$
$a + a = 2a$

NO NUMBER IN FRONT OF A LETTER? THEN AN INVISIBLE ONE IS ACTUALLY IN FRONT OF IT!

Any time you see a letter by itself, then there is really an invisible one in front of it

Add or subtract the following variables

1. $y + y + y = \boxed{3}y$

2. $6n - n = \boxed{5n}$

3. $12x + 2x - x = \boxed{13x}$

4. $590a + 10a - a = \boxed{599a}$

5. $r + r + 5r = \boxed{7r}$

NOW TRY THESE PROBLEMS!

12. $a + 3000a + 20a + 700a = \boxed{3721a}$

13. $500x + 1200x + 28x - x = \boxed{1727x}$

6. $4m + m + 3m = \boxed{8m}$

7. $12z + z + 2z = \boxed{15z}$

8. $20a - a + 20a - a = \boxed{38a}$

9. $15y + y + 15y + y = \boxed{32y}$

10.
$$\begin{array}{r} 15n \\ - n \\ \hline \boxed{14n} \end{array}$$

11. Rewrite the following problem with coefficients (numbers in front of letters)

$(n + n + n) - n - n$

$\boxed{} - \boxed{} = \boxed{}$

© Peter Wise, 2014

10

Adding Variables with Exponents

Examples

A. $3\boxed{a^2} + 4\boxed{a^2} = 7\boxed{a^2}$

Add the numbers
Copy the letters and exponents

RULE:
You can only add or subtract terms if they have the
- Same letter AND
- Same exponent

B.
$$\begin{array}{r} 6x^3 \\ + 4x^3 \\ \hline 10x^3 \end{array}$$

Add or subtract the following terms

AFTER YOU ADD THE COEFFICIENTS, JUST COPY ALL THE LETTERS AND EXPONENTS THEY WAY THEY ARE IN BOTH TERMS!

1. $4x^3 + 5x^3 = \boxed{9}x^3$

2. $2y^9 + 6y^9 = \boxed{8}y^{\boxed{9}}$

3. $2n^2 + 2n^2 + 2n^2 = \boxed{6}n^{\boxed{2}}$

4. $3a^2 + 4a^2 - 2a^2 = \boxed{5}a^2$

SUBTRACTION WORKS THE SAME WAY AS ADDITION!

5. $10x^2 + 2x^2 - 5x^2 = \boxed{7}x^{\boxed{2}}$

6. $10a^2 b^3 + 5a^2 b^3$
$= \boxed{15}a^{\boxed{2}}b^{\boxed{3}}$

7. $2x^5 y^2 z^4 + 3x^5 y^2 z^4$
$= \boxed{5}x^{\boxed{5}}y^{\boxed{2}}z^{\boxed{4}}$

8. $16a^7 b^3 - 2a^7 b^3 - 4a^7 b^3$
$= \boxed{10}a^{\boxed{7}}b^{\boxed{3}}$

© Peter Wise, 2014

11

Example

USE PARENTHESES WHERE EACH VARIABLE IS, THEN SUBSTITUTE THE NUMBER FOR THE VARIABLE AND CALCULATE!

A. ab $a = 2$
$a \cdot b$ $b = 5$
$(2)(5) = 10$

↑ substitute 2 for the letter a
↑ substitute 5 for the letter b

Substitute values for the variables and calculate

1. xy $x = 4$ $y = 3$
$(4)(3) = \boxed{12}$

REMEMBER! WHEN THEY TOUCH, THEY TIMES!

2. ab^2 $a = 2$ $b = 3$
$(2)(3^2) = \boxed{18}$

3. a^2b $a = 5$ $b = 2$
$(5^2)(2) = \boxed{50}$

4. a^2b^2 $a = 3$ $b = 10$
$(3^2)(10^2) = \boxed{900}$

write the exponents inside the parentheses

5. $\frac{a}{2} \cdot b$ $a = 8$ $b = 10$
$\frac{(8)}{2}(10) = \boxed{40}$

Order of Operations: multiply before you add

6. $xy + z$ $x = 4$ $y = 8$ $z = 2$
$(4 \cdot 8) + 2 = \boxed{34}$
$(32) + 2$

7. $\frac{m + n}{3}$ $m = 20$ $n = 7$
$\frac{(20) + (7)}{3} = \boxed{9}$

8. $r - st$ $r = 20$ $s = 4$ $t = 3$
$20 - (4 \cdot 3) = \boxed{8}$

© Peter Wise, 2014

12

Substitute values for the variables and calculate

1. $ab - cd$ $a = 4$ $b = 10$ $c = 3$ $d = 5$

TRADE THE LETTER A FOR 4!
TRADE THE LETTER B FOR 10!

$(4)(10) - (3)(5)$
$a = 4 \; b = 10 \quad c = 3 \; d = 5$
$\boxed{40} - \boxed{15} = \boxed{25}$

2. πr^2 $\pi = 3.14$ $r = 3$
$(3.14)(9)^2 = \boxed{28.26}$

Note: π is not a variable, it is a constant, meaning that it will always have a definite value; we'll use 3.14

3. $\frac{n^2}{2}$ $n = 4$
$\frac{16}{2} = \boxed{8}$

4. $3y^2$ $y = 10$
$(3)(100) = \boxed{300}$

Note:
Only the y is raised to the second power

5. $4ac$ $a = 3$ $c = 4$

PUT EVERYTHING IN SEPARATE PARENTHESES!

$(4)(3)(4) = \boxed{48}$

6. $b^2 - 4ac$ $a = 2$ $b = 8$ $c = 3$
$(64) - (4)(2)(3)$
$64 - 24 = 40$ $= \boxed{40}$

7. $\frac{a^2}{b} \cdot c$ $a = 6$ $b = 4$ $c = 3$
$\frac{36}{4} \cdot 3$
$9 \cdot 3 = 27$ $= \boxed{27}$

8. $a + bc$ $a = 12$ $b = 7$ $c = 8$
Watch for the order of operations!
$12 + (7)(8)$ $= \boxed{68}$
$12 + 56 = 68$

9. $5x^2 - xy$ $x = 3$ $y = 11$
$5(9) - (3)(11)$ $= \boxed{12}$
$45 - 33 = 12$

10. $r^2st - (r + s)$ $r = 3$ $s = 2$ $t = 4$
$(9)(2)(4) - (3 + 2)$ $= \boxed{67}$
$72 - 5 = 67$

© Peter Wise, 2014

13

Key words or phrases to look for:

"a number" "some number" → x (or any other variable letter)

"is" "the result is" → = (equal sign)

"sum" → addition
"increased by" → addition
"difference" → subtraction
"decreased by" → subtraction
"product" → multiplication
"quotient" → division

Watch out for these

"3 less than a number" → $x - 3$
watch the order! variable comes first

"3 more than a number" → $x + 3$

Translate the following words into algebraic symbols; solve them if you can!

Examples

A. The sum of 5 and a number is 25 $5 + x = 25$ $x = \boxed{20}$

B. The product of 4 and a number is 12 $4x = 12$ $x = \boxed{3}$

1. A number increased by 10 is 40 $x + 10 = 30$ $x = \boxed{30}$

2. 2 less than a number is 5 $x - 2 = 5$ $x = \boxed{7}$

3. The quotient of a number and 2 is 16 $x \div 2 = 16$ $x = \boxed{32}$

4. The difference of a number and 6 is 10 $x - 6 = 10$ $x = \boxed{16}$

© Peter Wise, 2014

14

Circle (a) or (b) to identify the equation that correctly represents the words

1. A number is increased by 3; the result is 20
(a) ⟨$x + 3 = 20$⟩
(b) $3x = 20$

2. The product of a number and 8 is 3 less than 35
(a) $8x = 3 - 35$
(b) ⟨$8x = 35 - 3$⟩

3. The product of a number and 5 equals the sum of the number and 16
(a) ⟨$5x = x + 16$⟩
(b) $5x + 16 = x$

4. The difference of 40 and a number equals the product of the (same) number and 3
(a) $x - 40 = 3x$
(b) ⟨$40 - x = 3x$⟩

5. The quotient of a number and 6 equals the (same) number decreased by 20
(a) $x \div 6 = 20 - x$
(b) ⟨$x \div 6 = x - 20$⟩

6. 6 less than a number equals the quotient of the number and 2
(a) ⟨$x - 6 = x \div 2$⟩
(b) $6 - x = x \div 2$

7. The sum of a number and 10 equals the difference of 20 and 2
(a) ⟨$x + 10 = 20 - 2$⟩
(b) $x + 10 = 20 \div 2$

8. The product of a number and 3 is increased by 8. The result is 20.
(a) $3x \cdot 8 = 20$
(b) ⟨$3x + 8 = 20$⟩

9. 5 less than a number is equal to the quotient of 30 and 3
(a) $5 - x = 30 \div 3$
(b) ⟨$x - 5 = 30 \div 3$⟩

© Peter Wise, 2014

15

More Words into Algebra Symbols

More Key words or phrases to look for:

"doubled"	→	multiplied by 2
"tripled"	→	multiplied by 3
"halved"	→	divided by 2
"squared"	→	raised to the 2nd power
"cubed"	→	raised to the 3rd power

Circle (a) or (b) to identify the equation that correctly represents the words

1. A number is doubled. The result is equal to half of 24.
   (a) $2x = 24 \cdot 2$
   (b) $2x = 24 \div 2$ ⟵ circled

2. A number squared equals the sum of the (same) number and 30
   (a) $x^2 = x + 30$ ⟵ circled
   (b) $x^2 + x = 30$

3. Triple a number equals the (same) number squared, then decreased by the (same) number
   (a) $x^3 = x - x^2$
   (b) $3x = x^2 - x$ ⟵ circled

4. A number cubed, then increased by 3 equals the product of the number and 10
   (a) $x^3 + 3 = 10x$ ⟵ circled
   (b) $x^3 + 10 = 3x$

5. The difference of a 20 and a number is equal to the (same) number squared
   (a) $20 - x = x^2$ ⟵ circled
   (b) $20 - x = 2x$

6. Half of a number is equal to 2 less than 20
   (a) $x \div 2 = 20 - 2$ ⟵ circled
   (b) $x \div 2 = 20 \div 2$

© Peter Wise, 2014

Turning Words into Algebra Symbols

Translate the following words into algebraic symbols

You don't need to solve these equations; just write them for now

1. Triple a number is 15 — $3x = 15$
2. Double a number; decrease it by 4; the result is 20 — $2x - 4 = 20$
3. 5 less than a number is 30 — $x - 5 = 30$
   watch the order on these!
4. 5 more than a number is 45 — $x + 5 = 45$
5. A number is increased by 4, then squared, the result is 144 — $(x + 4)^2 = 144$
   HINT! YOU'LL NEED TO USE PARENTHESES!
6. A number squared, then increased by 3, is 12 — $x^2 + 3 = 12$
7. A number squared equals 5 times the number, decreased by 4 (same variable) — $x^2 = 5x - 4$
8. A number tripled equals the number increased by 14 — $3x = x + 14$
9. Some number is doubled; it equals 24 decreased by that number — $2x = 24 - x$
10. A number doubled, then increased by 2 equals the number tripled, then decreased by 7 — $2x + 2 = 3x - 7$

© Peter Wise, 2014

Like Terms

18

- Matching Letters
- Matching Exponents (on the letters)

IT DOESN'T MATTER WHAT THE COEFFICIENTS ARE!

Examples

A. $2x \quad x$ — These are like terms
   $x^3 \quad 2x^2$ — These are NOT like terms

B. These are all like terms
   $3x^5y^2 \quad 3y^2x^5 \quad 2yyx^5$
   THE ORDER DOESN'T MATTER! *YY IS THE SAME AS Y-SQUARED!*

Circle the LIKE TERMS in each row

1. (x) (3x) x^2
2. (4x³) 4x (x³)
3. (5xy) $8x^2y^2$ (7yx)
4. $7a^3b^4$ (2a⁵b⁶) (9a⁵b⁶)
5. $2x^2y^3$ (3x³y²) (6y²x³) (2x³yy)
6. (7²c⁴d⁵) $7c^3d^5$ (7ccccd⁵)
7. $8x^7y^4z^3$ (5x⁷y³z⁴) (x⁷y³z⁴)
8. $3n^3x^2y^4$ (4n⁵x²y³) (.2)n⁵x²y³)

Add the LIKE TERMS

Example

A. $3a^3 + 4a^3 + 2a^5 = 7a^3 + 2a^5$
   CAN'T BE ADDED WITH THE OTHERS BECAUSE IT'S NOT A LIKE TERM!

1. $5x^2 + 4y^3 + 3x^2 = $ $8x^2$ + $4y^3$
   THE TERM YOU CAN'T ADD PUT HERE!

2. $2a^4b^3c^2 + 8a^5b^4c^3 + 6a^5b^4c^3 = $ $14a^5b^4c^3$ + $2a^4b^3c^2$

© Peter Wise, 2014

Like Terms

19

LOOK ONLY AT THE LETTERS AND THEIR EXPONENTS!   *COEFFICIENTS DON'T MATTER!*

LIKE TERMS:
- Matching letter(s) - order doesn't matter
- Matching exponent(s) - same exponents on the same letters

Like terms can have different coefficients

Examples

A. $a^2 \quad a^2$ — Like terms
   THE LETTERS AND THEIR EXPONENTS MATCH!

B. $x^2y^3 \quad x^2y^3$ — Like terms
   THE EXPONENT ON EACH X IS THE SAME; THE EXPONENT ON EACH Y IS THE SAME!

C. $3a^4b^5 \quad 7b^5a^4$ — Like terms
   THE ORDER OF THE LETTERS DOESN'T MATTER!

D. $5n^3m^4 \quad 5n^3$ — NOT like terms
   M⁴ IS MISSING!

E. $2r^2s^7 \quad 2r^3s^7$ — NOT like terms
   THE EXPONENT ON THIS R IS DIFFERENT FROM THE EXPONENT ON THE OTHER R!

Circle Y or N for Yes or No

1. $2x^3y^4 \quad 3x^3y^4$
   a) Same letters? (Y) N
   b) Do the same letters have the same exponents? (Y) N
   c) Are these LIKE TERMS? (Y) N

2. $3a^4b^5 \quad 7b^5a^4$ *ORDER DOESN'T MATTER!*
   a) Same letters? (Y) N
   b) Do the same letters have the same exponents? (Y) N
   c) Are these LIKE TERMS? (Y) N

3. $5n^7m^9 \quad 5n^8m^9$
   a) Same letters? (Y) N
   b) Do the same letters have the same exponents? Y (N)
   c) Are these LIKE TERMS? Y (N)

4. $4yyy \quad 9y^3$ *YYY = Y³!*
   a) Same letters? (Y) N
   b) Do the same letters have the same exponents? (Y) N
   c) Are these LIKE TERMS? (Y) N

© Peter Wise, 2014

Like Terms

- You can only ADD or SUBTRACT numbers or variables if they are LIKE TERMS
- If they are NOT like terms, keep them separate (don't add the coefficients)

Examples

A.

LIKE TERMS
You can add these

The exponent on the a doesn't match the other exponents on a Keep this one separate!

$3a^2 + 4a^2$ + $5a^3$ =

$7a^2$ + $5a^3$

YOU CAN'T ADD THE 7 AND THE 5 BECAUSE THE EXPONENTS DON'T MATCH!

Answer: $7a^2 + 5a^3$

Add or subtract the following terms

1. $4x^5 + 6x^5 + 3x^2 = $ $10x^5$ + $3x^2$

2. $7y^3 + 8y^2 + 2y^3 = $ $9y^3$ + $8y^2$

LOOK CLOSELY... TERMS THAT CANNOT BE COMBINED WILL HAVE TO BE ADDED OR SUBTRACTED AS THEY ARE!

3. $12n^4 - 2n^4 + 2y^3 = $ $10n^4$ + $2y^3$

4. $3x^2y^3 + 5x^2y^4 + 4x^2y^3 = $ $7x^2y^3$ + $5x^2y^4$

Look closely...
One term has to be separate

5. $15r^3s^5 - 2r^2s^2 - 3r^3s^5 = $ $12r^3s^5$ - $2r^2s^2$

6. $2a^2b^3 + 5x^3y^4 + 6a^2b^3 + 2x^3y^4 = $ $8a^2b^3$ + $7x^3y^4$

© Peter Wise, 2014

20

Undoing Addition and Subtraction

Goal: Isolate the variable (x = some number)

How to do this:
Undo addition by SUBTRACTION
Undo subtraction by ADDITION

Examples

3 IS BEING ADDED TO THE VARIABLE!

A.
$a + 3 = 10$
$-3 \quad -3$
$a = 7$

TO ISOLATE THE VARIABLE, UNDO +3, BY SUBTRACTING 3!

BUT THE RULE IS "WHATEVER YOU DO TO ONE SIDE...YOU HAVE TO DO TO THE OTHER!"

B.
$a - 2 = 6$
$+2 \quad +2$
$a = 8$

IN THIS CASE, UNDO MINUS 2 BY ADDING 2 TO BOTH SIDES!

Solve the following equations by reversing the addition/subtraction to the variable

1. $a + 5 = 20$
$-5 \quad -5$
$a = 15$

THESE SIGNS WILL ALWAYS BE OPPOSITE!

2. $a + 7 = 30$
$-7 \quad -7$
$a = 23$

3. $n - 5 = 12$
$+5 \quad +5$
$n = 17$

4. $6 + m = 15$
$-6 \quad -6$
$m = 9$

5. $c - 4 = 27$
$+4 \quad +4$
$c = 31$

6. $y - 4 = 21$
$+4 \quad +4$
$y = 25$

7. $-8 + x = 22$
$+8 \quad +8$
$x = 30$

8. $32 = 12 + n$
$-12 \quad -12$
$20 = n$

DO THIS BOX FIRST!

9. $k - 9 = 34$
$+9 \quad +9$
$k = 43$

© Peter Wise, 2014

21

Undoing Addition and Subtraction

Solve the following equations by reversing the addition/subtraction to the variable

1. $p - 6 = 13$
$+6 \quad +6$
$p = 19$

THESE SIGNS WILL ALWAYS BE OPPOSITE!

2. $n - 8 = 28$
$+8 \quad +8$
$n = 36$

3. $h + 4 = 32$
$-4 \quad -4$
$h = 28$

4. $9 + b = 20$
$-9 \quad -9$
$b = 11$

5. $y - 12 = 36$
$+12 \quad +12$
$y = 48$

6. $m - 15 = 46$
$+15 \quad +15$
$m = 61$

7. $14 + s = 32$
$-14 \quad -14$
$s = 18$

8. $v - 10 = 49$
$+10 \quad +10$
$v = 59$

9. $x + 18 = 62$
$-18 \quad -18$
$x = 44$

10. $81 = 17 + a$
$-17 \quad -17$
$64 = a$

ELIMINATE THE NUMBER ON THE SAME SIDE AS THE VARIABLE!

11. $45 = j - 17$
$+17 \quad +17$
$j = 62$

DO THIS BOX FIRST!

12. $m - 12 = 79$
$+12 \quad +12$
$m = 93$

13. $41 = 13 + y$
$-13 \quad -13$
$28 = y$

DO THIS BOX FIRST!

14. $-12 + r = 38$
$+12 \quad +12$
$r = 50$

15. $q + 17 = 21$
$-17 \quad -17$
$q = 4$

© Peter Wise, 2014

22

Undoing Multiplication & Division

Goal: Isolate the variable (like x = some number)

How to do this:
Undo multiplication by DIVISION
Undo division by MULTIPLICATION

A NUMBER IN THE DENOMINATOR IS JUST ANOTHER WAY TO SHOW DIVISION!

Examples

A.
$5a = 45$
$\div 5 \quad \div 5$
$a = 9$

B.
$a \div 3 = 4$
$\times 3 \quad \times 3$
$a = 12$

C.
$\frac{a}{2} = 7$
$\times 2 \quad \times 2$
$a = 14$

Solve the following equations by reversing the multiplication/division to the variable

WHEN THEY TOUCH, THEY TIMES!

1. $3a = 15$
$\div 3 \quad \div 3$
$a = 5$

THESE OPERATIONS WILL ALWAYS BE OPPOSITE!

2. $a \div 7 = 3$
$\times 7 \quad \times 7$
$a = 21$

3. $d \times 6 = 42$
$\div 6 \quad \div 6$
$d = 7$

4. $\frac{a}{6} = 3$
$\times 6 \quad \times 6$
$a = 18$

5. $4s = 48$
$\div 4 \quad \div 4$
$s = 12$

6. $n \div 8 = 7$
$\times 8 \quad \times 8$
$n = 56$

7. $9y = 54$
$\div 9 \quad \div 9$
$y = 6$

8. $24 = 8m$
$\div 8 \quad \div 8$
$m = 3$

9. $\frac{a}{12} = 4$
$\times 12 \quad \times 12$
$a = 48$

© Peter Wise, 2014

23

Undoing Multiplication & Division

Solve the following equations by reversing the addition/subtraction to the variable

1. $3a = 33$
$\div 3 \quad \div 3$
$a = 11$

2. $7y = 28$
$\div 7 \quad \div 7$
$y = 4$

3. $n \div 8 = 2$
$\times 8 \quad \times 8$
$n = 16$

4. $\frac{m}{5} = 2$
$\times 5 \quad \times 5$
$m = 10$

5. $\frac{y}{9} = 3$
$\times 9 \quad \times 9$
$y = 27$

6. $\frac{x}{3} = 7$
$\times 3 \quad \times 3$
$x = 21$

7. $48 = 8n$
$\div 8 \quad \div 8$ — *DO THIS BOX FIRST!*
$n = 6$

8. $n \div 5 = 7$
$\times 5 \quad \times 5$
$n = 35$

9. $9r = 54$
$\div 9 \quad \div 9$
$r = 6$

10. $\frac{b}{7} = 7$
$\times 7 \quad \times 7$
$b = 49$

11. $72 = 6c$
$\div 6 \quad \div 6$
$c = 12$

12. $n \div 7 = 8$
$\times 7 \quad \times 7$
$n = 56$

13. $72 = 9x$
$\div 9 \quad \div 9$ — *DO THIS BOX FIRST!*
$x = 8$

14. $\frac{n}{5} = 8$
$\times 5 \quad \times 5$
$n = 40$

15. $k \div 12 = 8$
$\times 12 \quad \times 12$
$k = 96$

24

Do the Algebra Steps

"WHATEVER YOU DO TO ONE SIDE, YOU HAVE TO DO TO THE OTHER!"

1. $3x + 2 = 29$ — (-2) (-2)
"CHANGE ONE THING... COPY AGAIN!"
FIRST, YOU WANT TO GET RID OF ANYTHING ADDED OR SUBTRACTED TO THE VARIABLE!
$3x = 27$
$\div 3 \quad \div 3$
WHAT DO YOU DIVIDE BY TO UNDO "TIMES 3"?
"CHANGE ONE THING... COPY AGAIN!"
NOW, YOUR ANSWER: $x = 3$

2. $4x + 5 = 29$ — (-5) (-5)
STEP ONE, GET RID OF ANYTHING ADDED OR SUBTRACTED!
$4x = 24$
$\div 4 \quad \div 4$
THIS IS STEP TWO, WHEN YOU DIVIDE TO UNDO THE MULTIPLICATION!
$x = 6$
HERE IS YOUR ANSWER!

3. $6x - 3 = 63$ — (+3) (+3)
HOW DO YOU UNDO MINUS?
STEP ONE: ANSWER GOES HERE!
$6x = 66$
$\div 6 \quad \div 6$
STEP TWO: DIVIDE TO UNDO THE MULTIPLICATION!
$x = 11$ — *ANSWER!*

REMEMBER: THE STEPS ARE JUST AS IMPORTANT AS THE ANSWER!

4. $7x - 6 = 50$ — (+6) (+6)
$7x = 56$
$\div 7 \quad \div 7$
$x = 8$

5. $8y + 2 = 34$ — (-2) (-2)
$8y = 32$
$\div 8 \quad \div 8$
$x = 4$

6. $4a - 8 = 28$ — (+8) (+8)
$4a = 36$
$\div 4 \quad \div 4$
$a = 9$

LEARN GOOD ALGEBRA TECHNIQUE! IN THE FUTURE YOU WILL GET PROBLEMS WITH TOO MANY STEPS TO DO MENTALLY!

25

Solving 2-Step Equations

1. $7a + 3 = 17$
YOU WANT THE + 3 TO DISAPPEAR!
"WHATEVER YOU DO TO ONE SIDE, YOU HAVE TO DO TO THE OTHER!"
$-3 \quad -3$
$7a = 14$ — *"CHANGE ONE THING... COPY AGAIN!"*
$\div 7 \quad \div 7$
UNDO "7 TIMES A" — $a = 2$

2. $5n + 6 = 21$
YOU WANT THE + 6 TO DISAPPEAR!
$-6 \quad -6$
REWRITE THE NEW EQUATION! — $5n = 15$
$\div 5 \quad \div 5$
UNDO "5 X N" — $n = 3$

3. $3y - 2 = 34$
$+2 \quad +2$
$3y = 36$
$\div 3 \quad \div 3$
$y = 12$

4. $6c - 8 = 34$
$+8 \quad +8$
$6c = 42$
$\div 6 \quad \div 6$
$c = 7$

5. $4y + 2 = -26$
$-2 \quad -2$
$4y = -28$
$\div 4 \quad \div 4$
$y = -7$

6. $6r + 12 = 60$
$-12 \quad -12$
$6r = 48$
$\div 6 \quad \div 6$
$r = 8$

26

Solving 2-Step Equations

1. $3x - 5 = 22$
GET RID OF THIS FIRST!
$+5 \quad +5$
$3x = 27$ — *REWRITE THE NEW EQUATION!*
$\div 3 \quad \div 3$
$x = 9$

2. $7a + 10 = 52$
$-10 \quad -10$
$7a = 42$
$\div 7 \quad \div 7$
$a = 6$

3. $5c + 5 = -45$
$-5 \quad -5$
$5c = -50$
$\div 5 \quad \div 5$
$c = -10$

4. $4n + 8 = 52$
$-8 \quad -8$
$4n = 44$
$\div 4 \quad \div 4$
$n = 11$

5. $8x - 5 = 59$
$+5 \quad +5$
$8x = 64$
$\div 8 \quad \div 8$
$x = 8$

6. $7y - 9 = 75$
$+9 \quad +9$
$7y = 84$
$\div 7 \quad \div 7$
$y = 12$

27

Exponents – Number of Repeated Factors

Example

A. x^4 (exponent) — A factor is a multiplied number

x is a factor four times
(base) (exponent)

$x \cdot x \cdot x \cdot x$

→ x (base) is a factor 4 (exponent) times → $x \cdot x \cdot x \cdot x$ (show the multiplication)

Error alert: It does not mean x times 4

Use the example above as a pattern for the following problems

Error alert: Do not multiply the base times the exponent

1. $y^2 \rightarrow$ y (base) is a factor 2 (exponent) times → $y \cdot y$ (show the multiplication)

2. $a^4 \rightarrow$ a (base) is a factor 4 (exponent) times → $a \cdot a \cdot a \cdot a$ (show the multiplication)

3. $n^3 \rightarrow$ n is a factor 3 times → $n \cdot n \cdot n$ (show the multiplication)

4. $r^5 \rightarrow$ r is a factor 5 times → $r \cdot r \cdot r \cdot r \cdot r$ (show the multiplication)

5. $m^3 \rightarrow$ m is a factor 3 times → $m \cdot m \cdot m$ (show the multiplication)

6. Show with an exponent: 3 is a factor 4 times → 3^4 → $3 \cdot 3 \cdot 3 \cdot 3$

7. Show with an exponent: 2 is a factor 5 times → 2^5 → $2 \cdot 2 \cdot 2 \cdot 2 \cdot 2$

8. Show with an exponent: 10 is a factor 3 times → 10^3 → $10 \cdot 10 \cdot 10$ (show the multiplication)

© Peter Wise, 2014

28

Variables with Exponents

Examples

A. $xx = x^2$ — 2 X'S ARE MULTIPLYING EACH OTHER!

B. $(x + 2)(x + 2) = (x + 2)^2$ — (X + 2) IS A FACTOR TWO TIMES!

Write the following terms with exponents

1. $yy = y^2$

2. $x^5 = x \cdot x \cdot x \cdot x \cdot x$

3. $rrr = r^3$

4. $nnnn = n^4$

5. $y^3 = y \cdot y \cdot y$

6. $mmm = \square^{\square}$

7. $d^4 = d \cdot d \cdot d \cdot d$

8. $(xyz)(xyz)(xyz) = (xyz)^3$

9. $(aa)(aa) = (aa)^2$ or a^4

10. $(ab)(ab) = (ab)^2$

WHEN TERMS INSIDE PARENTHESES ARE IDENTICAL AND MULTIPLIED 1 OR MORE TIMES, THEY CAN ALSO BE WRITTEN WITH EXPONENTS!

11. $(x + 2)(x + 2) = (x + 2)^2$

12. $(y + 3)(y + 3)(y + 3) = (y + 3)^3$

13. $(a + 2y)^2 = (a + 2y)(a + 2y)$

14. $(2 + m)^3 = (2 + m)(2 + m)(2 + m)$

15. $(xy)^3 = (xy)(xy)(xy)$

© Peter Wise, 2014

29

Multiplying Numbers and Letters

Examples

A. $a \cdot a = a^2$ — TWO A'S ARE MULTIPLYING EACH OTHER!

IF NO EXPONENT IS WRITTEN, IT'S AN INVISIBLE ONE!

B. $a^1 \cdot a^1 = a^2$

C. $a^2 \cdot a^3 = a^{2+3}$ or a^5

D. $3a \cdot 2a = 6a^2$ — $3 \cdot 2 = 6$; $a \cdot a = a$ squared

E. $2a^2 \cdot 5a^4 = 10a^6$

Multiply the numbers
Multiply the letters

- Whenever you multiply you add factors (= why you add exponents here)
- You are just adding numbers that are already being multiplied

---

Use the example above as a pattern for the following problems

1. $d \cdot d = d^2$

2. $r^2 \cdot r^2 = r^4$

3. $t^3 \cdot t^7 = t^{10}$

4. $m^2 \cdot m^4 \cdot m = m^7$ — IF YOU DON'T SEE AN EXPONENT ON A NUMBER OR A LETTER—IT'S REALLY AN INVISIBLE 1!

5. $x^7 \cdot x^7 = x^{14}$

6. $3x \cdot 4x = 12x^2$

7. $5y^2 \cdot 2y = 10y^3$ — HOW MANY Y'S ARE MULTIPLYING EACH OTHER?

8. $6a^3 \cdot 3a^4 = 18a^7$

9. $2x^{10} \cdot 8x^5 = 16x^{15}$

10. $7y^3 \cdot 4y^2 = 28y^5$

© Peter Wise, 2014

30

Multiplying Numbers and Letters

Example

A. $3a^2 \cdot 5a^{10} = 15a^{12}$

#1 Multiply the numbers $3 \cdot 5 = 15$

#2 Multiply the variables $a^2 \cdot a^{10} = a^{12}$ — EXPONENTS SHOW MULTIPLICATION!

Since the letters and numbers are multiplying each other, put them together ("when they touch, they times!") $15a^{12}$

Multiply the following terms

1. $2x \cdot 7x = 14x^2$

2. $2x^3 \cdot 7x^5 = 14x^8$

3. $4a^5 \cdot 6a^6 = 24a^{11}$

4. $8y^2 \cdot 2y^5 = 16y^7$

5. $12a^7 \cdot 3a^8 = 36a^{15}$

6. $3n \cdot 4n^2 = 12n^3$

7. $5r^3 \cdot 7r^9 = 35r^{12}$

8. $3a^2b^3 \cdot 2a^5b^5 = 6a^7b^8$

9. $6x^5y^6 \cdot 8x^4y^6 = 48x^9y^{12}$

10. $-4m^8n \cdot 7m^2n^4 = -28m^{10}n^5$

11. $6x^3y^3 \cdot 3x^4y^8 = 18x^7y^{11}$

12. $4r^4s^7 \cdot -8r^5s^9 = -32r^9s^{16}$

13. $2a^2 \cdot 3a^3 \cdot 5a^4 = 30a^9$

14. $-6x^4y^3 \cdot -9y^5x^2 = 54x^6y^8$ — WATCH THE ORDER ON THIS ONE!

© Peter Wise, 2014

31

Adding vs. Multiplying Variables

Examples

A. $a \cdot a = a^2$

B. $a^2 \cdot a^2 = a^4$

C. $\boxed{3}a^2 \cdot \boxed{2}a^5 = \boxed{6}a^7$

D. $a + a = 2a$

E. $a^3 + a^3 = 2a^3$

F. $3a^5 + 3a^5 = 6a^5$

Keep the variable and exponent the same (when you add them they all have to match)

Add the coefficients (numbers in front of the variable)

Follow the correct pattern to add or multiply the following expressions

1. $x + x = \boxed{2x}$

2. $(x)(x) = \boxed{x^2}$
   When parentheses touch, they times!

3. $3x + 3x = \boxed{6}x$

4. $3x \cdot 3x = \boxed{9}x^{\boxed{2}}$
   Remember to multiply both the numbers and the letters!

5. $4y \cdot 2y^2 = \boxed{8}y^{\boxed{3}}$

6. $5a^3 + 2a^3 = \boxed{7}a^{\boxed{3}}$

7. $5a^3 \cdot 2a^3 = \boxed{10}a^{\boxed{6}}$

8. $2y^5 + 2y^5 + 2y^5 = \boxed{6}y^{\boxed{5}}$
   HOW MANY y⁵S DO YOU HAVE?

9. $2x^7 \cdot 5x^3 \cdot 7x^5 = \boxed{70}x^{\boxed{15}}$

10. $(4m^3)(6m^4) = \boxed{24}m^{\boxed{7}}$

© Peter Wise, 2014

32

Dividing Same Bases

Examples

A. $x^5 \div x^2 = x^{5-2} = x^3$

B. $x^a \div x^b = x^{a-b}$

$\dfrac{x^5}{x^2} = \dfrac{\cancel{x}\cancel{x}xxx}{\cancel{x}\cancel{x}}$

THE BOTTOM FACTORS SUBTRACT FROM THE TOP FACTORS!

After you cancel two x's from both top and bottom 3 x's remain

C. $\boxed{10}x^7 \div \boxed{2}x^4 = \boxed{5}x^3$

· Divide the coefficients
· Divide the letters by subtracting exponents

• Whenever you divide you subtract factors (= why you subtract exponents here)

• When dividing you are removing factors

NOTE: YOU CAN ONLY ADD OR SUBTRACT EXPONENTS IF THE BASES ARE THE SAME!

Divide the following (remember, when you DIVIDE you are REMOVING factors)

1. $a^{10} \div a^2 = a^{\boxed{8}}$
   WITH COEFFICIENTS JUST DIVIDE NORMALLY: 12 ÷ 3!

2. $12a^{10} \div 3a^2 = \boxed{4}a^{\boxed{8}}$

3. $\dfrac{y^7}{y^5} = y^{\boxed{2}}$
   REMEMBER! DIVIDE BY THE DENOMINATOR!

4. $x^8y^6 \div x^4y^5 = x^{\boxed{4}}y^{\boxed{1}}$

5. $\dfrac{r^9s^7}{r^4s^5} = \boxed{r^5s^2}$

6. $\dfrac{15m^6}{5m^2} = \boxed{3m^4}$

7. $18s^{12}t^{15} \div 3s^9t^8 = \boxed{6s^3t^7}$

8. $2a^3 \cdot 6a^5 \div 4a^2 = \boxed{3a^6}$

9. $\dfrac{24y^9}{3y^4} \cdot 2y^2 = \boxed{16y^7}$

10. $\dfrac{35a^{12}}{7a^{10}} \cdot 4a^3 = \boxed{20a^5}$

33

Exponent Practice

1. $r \cdot r = r^{\boxed{2}}$

2. $a^5 \div a^2 = a^{\boxed{3}}$

3. $y^2 \cdot y^2 \cdot y^2 = y^{\boxed{6}}$

4. $(x-3)(x-3)(x-3) = (x-3)^{\boxed{3}}$

5. $k^2 \cdot k^{10} \div k^3 = k^{\boxed{9}}$

6. $b^5 \div b = b^{\boxed{4}}$

7. $a^5 \cdot a^5 \cdot a^5 = a^{\boxed{15}}$

8. $\dfrac{d^7}{d^5} = d^{\boxed{2}}$

9. $\dfrac{n^7 \cdot n^3}{n^2} = n^{\boxed{8}}$

10. $(4x+y)(4x+y)(4x+y) = \boxed{(4x+y)}^{\boxed{3}}$

11. $yyyy = y^{\boxed{4}}$

12. $\dfrac{yyyyy}{yy} = y^{\boxed{3}}$

13. $3x^3 \cdot 3x^3 = \boxed{9}x^{\boxed{6}}$

14. $2x^5 \cdot 6x^2 = \boxed{12}x^{\boxed{7}}$

15. $15x^9 \div 5x^4 = \boxed{3}x^{\boxed{5}}$

16. $2x^4 \cdot 2x^4 \cdot 2x^4 = \boxed{8}x^{\boxed{12}}$

17. $3y^5 \cdot 6y^2 \div 2y^3 = \boxed{9}y^{\boxed{4}}$

18. $2x^2y^3 \cdot 3x^6y^7 = \boxed{6}x^{\boxed{8}}y^{\boxed{10}}$

19. $18a^8b^{10} \div 9a^3b^2 = \boxed{2}a^{\boxed{5}}b^{\boxed{8}}$

© Peter Wise, 2014

34

Equation Practice

GET RID OF THIS FIRST!

1. $6x - 5 = 19$
   $+\boxed{5} \quad +\boxed{5}$
   $\boxed{6x} = \boxed{24}$ REWRITE THE NEW EQUATION!
   $\div\boxed{6} \quad \div\boxed{6}$
   $\boxed{x = 4}$

2. $4y + 3 = 35$
   $-\boxed{3} \quad -\boxed{3}$
   $\boxed{4y} = \boxed{32}$
   $\div\boxed{4} \quad \div\boxed{4}$
   $\boxed{y = 8}$

3. $7a - 8 = -29$
   $+\boxed{8} \quad +\boxed{8}$
   $\boxed{7a} = \boxed{-21}$
   $\div\boxed{7} \quad \div\boxed{7}$
   $\boxed{a = -3}$

4. $5b - 3 = 42$
   $+\boxed{3} \quad +\boxed{3}$
   $\boxed{5b} = \boxed{45}$
   $\div\boxed{5} \quad \div\boxed{5}$
   $\boxed{b = 9}$

5. $9x + 5 = -58$
   $-\boxed{5} \quad -\boxed{5}$
   $\boxed{9x} = \boxed{-63}$
   $\div\boxed{9} \quad \div\boxed{9}$
   $\boxed{x = -7}$

6. $8c + 4 = 100$
   $-\boxed{4} \quad -\boxed{4}$
   $\boxed{8c} = \boxed{96}$
   $\div\boxed{8} \quad \div\boxed{8}$
   $\boxed{c = 12}$

© Peter Wise, 2014

35

Equation Practice

Combine like terms and solve for the variable

1. $2x = 15 + 3$
$2x = \boxed{18}$
$\boxed{÷2} \quad \boxed{÷2}$
$\boxed{x = 9}$

2. $7 + 8 = 5x$
$\boxed{15} = 5x$
$\boxed{÷5} \quad \boxed{÷5}$
$\boxed{x = 3}$

3. $6x = 38 + 4$
$\boxed{6x} = \boxed{42}$
$\boxed{÷6} \quad \boxed{÷6}$
$\boxed{x = 7}$

4. $9x = 22 + 5$
$\boxed{9x = 27}$
$\boxed{÷9} \quad \boxed{÷9}$
$\boxed{x = 3}$

5. $27 + 3 = 6x$
$\boxed{30 = 6x}$
$\boxed{÷6} \quad \boxed{÷6}$
$\boxed{5 = x}$

DIVIDE BY THE COEFFICIENT OF X!

6. $7x = 30 + 26$
$\boxed{7x = 56}$
$\boxed{÷7} \quad \boxed{÷7}$
$\boxed{x = 8}$

7. $11x = 92 - 4$
$\boxed{11x = 88}$
$\boxed{÷11} \quad \boxed{÷11}$
$\boxed{x = 8}$

8. $39 + 9 = 8x$
$\boxed{48 = 8x}$
$\boxed{÷8} \quad \boxed{÷8}$
$\boxed{6 = x}$

9. $9x = 52 + 11$
$\boxed{9x = 63}$
$\boxed{÷9} \quad \boxed{÷9}$
$\boxed{x = 7}$

10. $12x = 31 - 7$
$\boxed{12x = 24}$
$\boxed{÷12} \quad \boxed{÷12}$
$\boxed{x = 2}$

11. $9x = 27 + 9$
$\boxed{9x = 36}$
$\boxed{÷9} \quad \boxed{÷9}$
$\boxed{x = 4}$

12. $25 - 8 = 17x$
$\boxed{17 = 17x}$
$\boxed{÷17} \quad \boxed{÷17}$
$\boxed{1 = x}$

© Peter Wise, 2014

36

Combine Variables; Then Solve

Example

A. $2x = 30 - 4x$ 〔+ 4x〕 〔+ 4x〕

WHEN YOU ADD 4X TO BOTH SIDES, YOU ADD IT TO 2X!

change sides, change signs

$\dfrac{6x}{6} = \dfrac{30}{6}$

$x = 5$

YOU COULD SUBTRACT 2X FROM BOTH SIDES...BUT IT IS EASIER TO SUBTRACT 4X FROM BOTH SIDES BECAUSE YOU'LL GET A POSITIVE VALUE FOR THE VARIABLE!

Get the letter(s) on one side

Get the number on the other side

Tip: It is usually easiest if you get the bigger (positive) value of the letter on one side and the number on the other side.

IT DOESN'T MATTER WHICH SIDE THE YOU PUT THE LETTER OR NUMBER ON

1. $-2x + 8x = 18$
$6x = 18$
$x = 3$

4. $x + 10 = -2 - 2x$
$3x = -12$
$x = -4$

2. $5x = 21 + 2x$
$3x = 21$
$x = 7$

5. $-3x = -7x + 8$
$4x = 8$
$x = 2$

3. $4x = -10 - 4x - 6$
$8x = -16$
$x = -2$

6. $36 - 2x = x$
$3x = 36$
$x = 12$

37

Equations: Combining Variables

Combine like terms and solve for the variable

1. $2x + 5x = 14$
$\boxed{7x} = 14$
$(÷)\boxed{7} \quad (÷)\boxed{7}$
$\boxed{x = 2}$

ONE WAY TO SHOW DIVISION IS BY PUTTING THE DIVISOR UNDER A FRACTION BAR!

2. $-8 - 8 = 12x - 4x$
$\boxed{-16} = \boxed{8x}$
$\boxed{8} \quad \boxed{8}$
$\boxed{x = -2}$

3. $18x - 15x = 17 + 16$
$\boxed{3x} = \boxed{33}$
$\boxed{3} \quad \boxed{3}$
$\boxed{x = 11}$

4. $-30 + 6 = 15x - 7x$
$\boxed{-24} = \boxed{8x}$
$\boxed{8} \quad \boxed{8}$
$\boxed{x = -3}$

5. $8x - 3x = -55 + 10$
$\boxed{5x} = \boxed{-45}$
$\boxed{5} \quad \boxed{5}$
$\boxed{x = -9}$

6. $17x - 15x = 12 + 14$
$\boxed{2x} = \boxed{26}$
$\boxed{2} \quad \boxed{2}$
$\boxed{x = 13}$

7. $14x - 8x = 22 + 26$
$\boxed{6x} = \boxed{48}$
$\boxed{6} \quad \boxed{6}$
$\boxed{x = 8}$

8. $13x - 4x = 70 - 7$
$\boxed{9x} = \boxed{63}$
$\boxed{9} \quad \boxed{9}$
$\boxed{x = 7}$

© Peter Wise, 2014

38

Vertical Angle Equations

Example

A. SINCE VERTICAL ANGLES ARE EQUAL, IT'S JUST LIKE BOTH ANGLES ARE TWO SIDES OF AN EQUATION!

$20°$
$2x$

$\dfrac{2x}{2} = \dfrac{20°}{2}$ $\boxed{x = 10°}$

Keep your eyes out for hidden equations like these

Since vertical angles are equal, find the measure of each angle

1. $2x$ $16°$
$\boxed{2x} = \boxed{16}° \quad x = \boxed{8}°$

2. $35°$ $7x$
$\boxed{7x} = \boxed{35}° \quad x = \boxed{5}°$

3. $90°$ $2x$
$\boxed{2x} = \boxed{90}° \quad x = \boxed{45}°$

4. $2x + 56°$ $10x$
$10x = 2x + 56°$ ← Set up your equation
$8x = 56$ ← Subtract from both sides
$x = 7$ ← Divide from both sides
$x = \boxed{7}°$

5. $-4x + 42°$ $3x$
$3x = -4x + 42°$
$7x = 42°$
$x = 6°$ $x = \boxed{6}°$

© Peter Wise, 2014

39

Equations with Complementary Angles

Complementary Angles Add up to 90°

Examples

A. $2x + 10°$ / $20°$

Add the values of both angles and set them equal to 90°

$(2x + 10°) + 20° = 90°$ ← Add the 10° and the 20°

$2x + 30° = 90°$ ← Subtract 30° from both sides

$2x = 60°$ ← Divide both sides by 2

$x = 30°$

B. $4x + 1°$ / $3x + 5°$

Add the values of both angles and set them equal to 90°

$(4x + 1°) + (3x + 5°) = 90°$ ← Add the degrees and the variables

$7x + 6° = 90°$ ← Subtract 6° from both sides

$7x = 84°$ ← Divide both sides by 7

$x = 12°$

Add both angles and set them equal to 90° to find the measure of x

1. $30°$ / $3x + 15°$

$30° + (3x + 15°) = 90°$ ← Set up your equation. Both angles together equal 90°

$3x + 45° = 90°$ ← Add the degrees on the left side of the equation

$3x = 45°$ ← Eliminate the degrees on the left side by subtracting

$x = 15°$ ← Divide both sides by the coefficient (number in front of the variable)

2. $2x + 30$ / $3x + 10$

$(2x + 30°) + (3x + 10°) = 90°$ ← Set up your equation. Both angles together equal 90°

$5x + 40° = 90°$ ← On the left side of the equation add both pairs of like terms

$5x = 50°$ ← Eliminate the degrees on the left side by subtracting

$x = 10°$ ← Divide both sides by the coefficient (number in front of the variable)

3. $x - 22°$ / $2x + 1°$

$(x - 22°) + (2x + 1°) = 90°$ ← Set up your equation. Both angles together equal 90°

$3x - 21° = 90°$ ← On the left side of the equation add both pairs of like terms

$3x = 111°$ ← Eliminate the degrees on the left side by adding

$x = 37°$ ← Divide both sides by the coefficient (number in front of the variable)

40

Equations with Supplementary Angles

Supplementary Angles Add up to 180°

Example

A. $2x + 20°$ / $3x + 10°$

Add the values of both angles and set them equal to 180°

$(2x + 20°) + (3x + 10°) = 180°$

$5x + 30° = 180°$ ← Combine the like terms

$5x = 150°$ ← Subtract 30° from both sides

$x = 30°$ ← Divide both sides by 5

Add both angles and set them equal to 180° to find the measure of x

1. $9x - 3°$ / $3x + 15°$

$(9x - 3°) + (3x + 15°) = 180°$ ← Set up your equation. Both angles together equal 180°

$12x + 12° = 180°$ ← Add the like terms on the left side of the equation

$12x = 168°$ ← Eliminate the degrees on the left side by subtracting

$x = 14°$ ← Divide both sides by the coefficient (number in front of the variable)

2. $7x + 47°$ / $23x + 43°$

$(7x + 47°) + (23x + 43°) = 180°$ ← Set up your equation. Both angles together equal 180°

$30x + 90° = 180°$ ← Add the like terms on the left side of the equation

$30x = 90°$ ← Eliminate the degrees on the right side by subtracting

$x = 3°$ ← Divide both sides by the coefficient (number in front of the variable)

3. $4x - 11°$ / $x + 91°$

$(4x - 11°) + (x + 91°) = 180°$ ← Set up your equation. Both angles together equal 180°

$5x + 80° = 180°$ ← Add the like terms on the left side of the equation

$5x = 100°$ ← Eliminate the degrees on the left side by subtracting

$x = 20°$ ← Divide both sides by the coefficient (number in front of the variable)

41

Algebra With Perimeters

1. (square with sides x, x, x, x)

What is the perimeter of this shape, in terms of x? (count around, add up all the variables)

$\boxed{4}$ x

REMEMBER THAT A LETTER BY ITSELF HAS AN INVISIBLE 1 IN FRONT OF IT!

If the measurement of the perimeter is 20 inches, what is the value of x?

$\boxed{4}$ x = $\boxed{20}$ in   answer: $\boxed{5}$ in

2. (rectangle with sides $2x$, x, $2x$, x)

What is the perimeter of this shape, in terms of x? (count around, add up all the variables)

$\boxed{6}$ x

If the measurement of the perimeter is 18 mm, what is the value of x?

$\boxed{6}$ x = $\boxed{18}$ mm   answer: $\boxed{3}$ mm

3. (triangle with sides $5x$, $5x$, $5x$)

(Equilateral triangle)

What is the perimeter of this shape, in terms of x? (count around, add up all the variables)

$\boxed{15}$ x

If the measurement of the perimeter is 30 cm, what is the value of x?

$\boxed{15}$ x = $\boxed{30}$ cm   answer: $\boxed{2}$ cm

42

Equation Practice

Example

A. $4x + 5 = 25$

Subtract from both sides $\boxed{-5 \quad -5}$

Copy the new equation $\boxed{4x = 20}$

Undo the multiplication (divide both sides) $\dfrac{4x}{4} \quad \dfrac{}{4}$

Write the value of x $\boxed{x = 5}$

3. $7x + 8 = 29$

$\boxed{-8 \quad -8}$

Copy the new equation $\boxed{7x = 21}$

$\boxed{7} \quad \boxed{7}$

$\boxed{x = 3}$

1. $6x - 4 = 20$

Add to both sides $\boxed{+4 \quad +4}$

Copy the new equation $\boxed{6x = 24}$

divide both sides $\boxed{6} \quad \boxed{6}$

Write the value of x $\boxed{x = 4}$

4. $3x - 3 = 24$

$\boxed{+3 \quad +3}$

Copy the new equation $\boxed{3x = 27}$

$\boxed{3} \quad \boxed{3}$

$\boxed{x = 9}$

2. $6x - 7 = 5$

$\boxed{+7 \quad +7}$

Copy the new equation $\boxed{6x = 12}$

$\boxed{6} \quad \boxed{6}$

$\boxed{x = 2}$

5. $52 = 4 + 8x$

$\boxed{-4 \quad -4}$

Copy the new equation $\boxed{48 = 8x}$

$\boxed{8} \quad \boxed{8}$

$\boxed{x = 6}$

43

Equation Practice

1. $9x + 6 = 78$

Subtract from both sides: $-6 \quad -6$

Copy the new equation: $9x = 72$

divide both sides: $\dfrac{9}{9} \quad \dfrac{9}{9}$

Write the value of x: $x = 8$

2. $30 = 8y - 10$

$+10 \quad +10$

Copy the new equation: $40 = 8y$

$\dfrac{8}{8} \quad \dfrac{8}{8}$

$y = 5$

3. $7s + 5 = 40$

$-5 \quad -5$

Copy the new equation: $7s = 35$

$\dfrac{7}{7} \quad \dfrac{7}{7}$

$s = 5$

4. $8g - 13 = 43$

$+13 \quad +13$

$8g = 56$

$\dfrac{8}{8} \quad \dfrac{8}{8}$

$g = 7$

5. $4x + 7 = 55$

$-7 \quad -7$

$4x = 48$

$\dfrac{4}{4} \quad \dfrac{4}{4}$

$x = 12$

6. $50 = 6a + 8$

$-8 \quad -8$

$42 = 6a$

$\dfrac{6}{6} \quad \dfrac{6}{6}$

$a = 7$

44

Equation Practice

Solve the following equations (do the algebra steps!)

1. $94 = 10 + 7s$

Subtract from both sides: $-10 \quad -10$

Copy the new equation: $84 = 7s$

divide both sides: $\boxed{7} \quad \boxed{7}$

Write the value of s: $s = 12$

2. $26 = 14 + 4x$

$-14 \quad -14$

$\dfrac{12}{4} = \dfrac{4x}{4}$

$x = 3$

3. $12c - 15 = 33$

$+15 \quad +15$

$\dfrac{12c}{12} = \dfrac{48}{12}$

$c = 4$

4. $7y + 17 = 80$

$-17 \quad -17$

$\dfrac{7y}{7} = \dfrac{63}{7}$

$y = 9$

5. $4n - 20 = 0$

$+20 \quad +20$

$\dfrac{4n}{4} = \dfrac{20}{4}$

$n = 5$

6. $73 = 11x + 7$

$-7 \quad -7$

$\dfrac{66}{11} = \dfrac{11x}{11}$

$x = 6$

45

Adding & Subtracting Variables as Numerators

Example

A. $\dfrac{3x}{y} + \dfrac{2x}{y} = \dfrac{5x}{y}$ ← Add or subtract numerators
← Copy denominators

These all must match (no matter what they are)

Add/subtract the following fractions

1. $\dfrac{17n}{r} + \dfrac{3n}{r} = \dfrac{20r}{r}$

2. $\dfrac{6a}{b} - \dfrac{4a}{b} = \dfrac{2a}{b}$

3. $\dfrac{16y}{m} + \dfrac{8y}{m} - \dfrac{2y}{m} = \dfrac{22y}{m}$

4. $\dfrac{20a}{c-d} + \dfrac{15a}{c-d} - \dfrac{6a}{c-d} = \dfrac{29a}{c-d}$

5. $\dfrac{38m}{y} - \dfrac{5m}{y} = \dfrac{33m}{y}$

6. $\dfrac{20a}{x^2 - y^2} + \dfrac{8a}{x^2 - y^2} = \dfrac{28a}{x^2 - y^2}$

46

Variables Divided by Numbers

Examples

A. $\dfrac{x}{3} = 2$

$(3)\dfrac{x}{3} = 2(3)$

THE WAY TO UNDO ÷ 3 IS TO MULTIPLY BY 3!

$x = 6$

B. $\dfrac{x}{5} + 2 = 5$

$\boxed{-2} \quad \boxed{-2}$

$\dfrac{x}{5} = 3$

$(5)\dfrac{x}{5} = 3(5)$

$\boxed{x = 15}$

Solve for x by undoing the division

1. $\dfrac{x}{4} = 2$

$(4)\dfrac{x}{4} = 2(4)$

MULTIPLYING BY THE SAME NUMBER CANCELS OUT THE DENOMINATOR!

$x = 8$

2. $\dfrac{x}{10} = 7$

$(10)\dfrac{x}{10} = 7(10)$

$x = 70$

3. $5 = \dfrac{x}{9}$

$(9)5 = \dfrac{x}{9}(9)$

$45 = x$

4. $\dfrac{x}{7} = 1 + 3$

$(7)\dfrac{x}{7} = \boxed{4}(7)$

$x = 28$

5. $6 = \dfrac{x}{2}$

$(2)6 = \dfrac{x}{2}(2)$

$12 = x$

6. $\dfrac{y}{8} = 3$

$(8)\dfrac{y}{8} = 3(8)$

$y = 24$

47

Making Fraction Coefficients Disappear

Examples

THIS FRACTION WILL DISAPPEAR WHEN YOU MULTIPLY IT BY ITS RECIPROCAL!

Rule
Fractions disappear (turn to 1) when you multiply them by their reciprocals

A. $\frac{2}{3}x = 6$

$\left(\frac{3}{2}\right)$ Multiply both sides of the equation by the reciprocal $\left(\frac{3}{2}\right)$

$\left(\frac{3}{2}\right)\frac{2}{3}x = 6\left(\frac{3}{2}\right)$ $x = 9$

answer

THESE FRACTIONS CANCEL EACH OTHER OUT!

When you multiply a fraction by its reciprocal you always get a new fraction equal to one

---

Solve the following equations by multiplying both sides by the fraction reciprocals

1. $\frac{2}{5}x = 8$

DIVIDE BY THE BOTTOM AND MULTIPLY BY THE TOP!

$\left(\frac{5}{2}\right)\frac{2}{5}x = 8\left(\frac{5}{2}\right)$

THE RECIPROCAL (FLIP) OF 2/5 GOES HERE!

$x = \boxed{20}$

3. $\frac{5}{9}x = 15$

$\left(\frac{9}{5}\right)\frac{5}{9}x = 15\left(\frac{9}{5}\right)$

$x = \boxed{27}$

2. $\frac{3}{4}x = 6$

$\left(\frac{4}{3}\right)\frac{3}{4}x = 6\left(\frac{4}{3}\right)$

$x = \boxed{8}$

4. $\frac{4}{3}x = 20$

$\left(\frac{3}{4}\right)\frac{4}{3}x = 20\left(\frac{3}{4}\right)$

$x = \boxed{15}$

48

Fraction Coefficients Continued

Solve the following equations by multiplying both sides by the fraction reciprocals

1. $\frac{7}{3}x = 21$

$\left(\frac{3}{7}\right)\frac{7}{3}x = 21\left(\frac{3}{7}\right)$

$x = \boxed{9}$

5. $\frac{8}{3}x = 16$

$\left(\frac{3}{8}\right)\frac{8}{3}x = 16\left(\frac{3}{8}\right)$

$x = \boxed{6}$

2. $\frac{5}{6}x = 30$

$\left(\frac{6}{5}\right)\frac{5}{6}x = 30\left(\frac{6}{5}\right)$

$x = \boxed{36}$

6. $\frac{6}{7}x = 18$

$\left(\frac{7}{6}\right)\frac{6}{7}x = 18\left(\frac{7}{6}\right)$

$x = \boxed{21}$

3. $\frac{3}{8}x = 12$

$\left(\frac{8}{3}\right)\frac{3}{8}x = 12\left(\frac{8}{3}\right)$

$x = \boxed{32}$

7. $3\frac{1}{3}x = 40$

$\left(\frac{3}{10}\right)\frac{10}{3}x = 40\left(\frac{3}{10}\right)$

$x = \boxed{12}$

4. $\frac{5}{4}x = 20$

$\left(\frac{4}{5}\right)\frac{5}{4}x = 20\left(\frac{4}{5}\right)$

$x = \boxed{16}$

8. $3\frac{3}{4}x = 30$

$\left(\frac{4}{15}\right)\frac{15}{4}x = 30\left(\frac{4}{15}\right)$

$x = \boxed{8}$

© Peter Wise, 2014

49

Fraction Coefficient Word Problems

Example

A. 2/3 of a number is 12. What is the number?

$\frac{2}{3}x = 12$

$\left(\frac{3}{2}\right)$ Multiply both sides of the equation by the reciprocal $\left(\frac{3}{2}\right)$ $x = \boxed{18}$

Write the following word problems as equations with fraction coefficients, then solve

1. 3/4 of a number is 9. What is the number?

$\boxed{\frac{3}{4}}\,x = \boxed{9}$ $x = \boxed{12}$

4. 7/9 of the weight is 14 lbs. What is the total weight?

$\boxed{\frac{7}{9}}\,x = \boxed{14}$ $x = \boxed{18}$ lbs

2. 5/7 of a number is 15. What is the number?

$\boxed{\frac{5}{7}}\,x = \boxed{15}$ $x = \boxed{21}$

5. 5/11 of the money is $250. What is the total amount of money?

$\frac{5}{11}x = \$250$ $x = \boxed{\$550}$

3. 4/5 of the distance is 20 miles. How far is the total distance?

$\boxed{\frac{4}{5}}\,x = \boxed{20}$ $x = \boxed{25}$ mi

6. 3/8 of the water is 300 gallons. What is the total amount?

$\frac{3}{8}x = 300$ gal $x = \boxed{800}$ gal

© Peter Wise, 2014

50

Equation Practice

GET X ON THE SIDE WHERE IT IS GREATER!

X IS GREATER ON THIS SIDE!

SO GET RID OF X ON THIS SIDE!

A. $10x = 3x + 21$

$-\boxed{3x} \quad -\boxed{3x}$

tip: Get x on one side and the numbers on the other side

$\boxed{7x} = \boxed{21}$

$\div\boxed{7} \quad \div\boxed{7}$

$\boxed{x} = \boxed{3}$

B. $3x = -2x + 10$

$+\boxed{2x} \quad +\boxed{2x}$

$\boxed{5x} = \boxed{10}$

$\div\boxed{5} \quad \div\boxed{5}$

$\boxed{x} = \boxed{2}$

Get x on the side where it's greater, then solve

1. $10x = 2x + 32$

$-\boxed{2x} \quad -\boxed{2x}$

$\boxed{8x} = \boxed{32}$

$\div\boxed{8} \quad \div\boxed{8}$

$x = \boxed{4}$

2. $36x = -4x + 80$

$+\boxed{4x} \quad +\boxed{4x}$

$\boxed{40x} = \boxed{80}$

$\div\boxed{40} \quad \div\boxed{40}$

$x = \boxed{2}$

3. $14x = 12x + 24$

$-\boxed{12x} \quad -\boxed{12x}$

$\boxed{2x} = \boxed{24}$

$\div\boxed{2} \quad \div\boxed{2}$

$x = \boxed{12}$

4. $7x = -2x + 27$

$+\boxed{2x} \quad +\boxed{2x}$

$\boxed{9x} = \boxed{27}$

$\div\boxed{9} \quad \div\boxed{9}$

$x = \boxed{3}$

5. $3x = -3x + (30 + 6)$

$+\boxed{3x} \quad +\boxed{3x}$

$\boxed{6x} = \boxed{36}$

$\div\boxed{6} \quad \div\boxed{6}$

$x = \boxed{6}$

6. $17x = 10x + 35$

$-\boxed{10x} \quad -\boxed{10x}$

$\boxed{7x} = \boxed{35}$

$\div\boxed{7} \quad \div\boxed{7}$

$x = \boxed{5}$

© Peter Wise, 2014

51

1. $4x = 30 + 2x$

$2x = 30$ $x = \boxed{15}$
$x = 15$

6. $10x - 6x - 70 = -8x + 2$

$12x = -72$ $x = \boxed{-6}$
$x = -6$

2. $-7x = 36 - 11x$

$4x = 36$ $x = \boxed{9}$
$x = 9$

7. $4x + 13 = -5 - 2x$

$6x = -18$ $x = \boxed{-3}$
$x = -3$

3. $2x - 8 = 40 - 4x$

$6x = 48$ $x = \boxed{8}$
$x = 8$

8. $-2x - 4 = -10x + 60$

$8x = 64$ $x = \boxed{8}$
$x = 8$

4. $4x - 70 = -3x + 7$

$7x = 77$ $x = \boxed{11}$
$x = 11$

9. $-3x + 30 = -10x - 5$

$7x = -35$ $x = \boxed{-5}$
$x = -5$

5. $-2x - 3 = -11x + 60$

$9x = 63$ $x = \boxed{7}$
$x = 7$

10. $5x - 40 = 8 - 7x$

$12x = 48$ $x = \boxed{4}$
$x = 4$

Example

A. $\frac{2}{9} = \frac{x}{12}$

#1 Cross multiply $9x = 24$

#2 Divide both sides by the number in front of the letter (the "coefficient") $\rightarrow \frac{9x}{9} \rightarrow \frac{24}{9}$

#3 Simplify $x = \frac{24}{9} = 2\frac{6}{9}$

$= \boxed{2\frac{2}{3}}$

Solve the following proportions; leave your answers as fractions or mixed numbers

1. $\frac{2}{7} = \frac{x}{3}$

#1 Cross multiply $\boxed{7x = 6}$

#2 Divide both sides by the number in front of the letter (the "coefficient")

$\frac{7x}{7} = \frac{6}{7}$ $x = \boxed{\frac{6}{7}}$

2. $\frac{4}{11} = \frac{x}{12}$

#1 Cross multiply

#2 Divide both sides $\frac{11x}{11} = \frac{48}{11}$

$x = \boxed{4\frac{4}{11}}$

3. $\frac{3}{7} = \frac{x}{10}$

#1
#2 $\frac{7x}{7} = \frac{30}{7}$ $x = \boxed{4\frac{2}{7}}$

4. $\frac{3}{5} = \frac{4}{x}$

$\frac{3x}{3} = \frac{20}{3}$

$x = \boxed{6\frac{2}{3}}$

5. $\frac{x}{5} = \frac{2}{3}$

$\frac{3x}{3} = \frac{10}{3}$

$x = \boxed{3\frac{1}{3}}$

Solve the following proportions; leave your answers as fractions or mixed numbers

1. $\frac{2}{5} = \frac{x}{8}$

#1 Cross multiply $\boxed{5x = 16}$

#2 Divide both sides by the number in front of the letter (the "coefficient")

$\frac{5x}{5} = \frac{16}{5}$ $x = \boxed{3\frac{1}{5}}$

2. $\frac{3}{4} = \frac{x}{7}$

#1 Cross multiply
#2 Divide both sides $\frac{4x}{4} = \frac{21}{4}$

$x = \boxed{5\frac{1}{4}}$

3. $\frac{2}{x} = \frac{7}{6}$

#1
#2 $\frac{7x}{7} = \frac{12}{7}$ $x = \boxed{1\frac{5}{7}}$

4. $\frac{x}{5} = \frac{7}{2}$

#1
#2 $\frac{2x}{2} = \frac{35}{2}$ $x = \boxed{17\frac{1}{2}}$

5. $\frac{7}{9} = \frac{5}{x}$

$\frac{7x}{7} = \frac{45}{7}$ $x = \boxed{6\frac{3}{7}}$

6. $\frac{4}{9} = \frac{x}{5}$

$\frac{9x}{9} = \frac{20}{9}$ $x = \boxed{2\frac{2}{9}}$

7. $\frac{2}{5} = \frac{x}{12}$

$\frac{5x}{5} = \frac{24}{5}$ $x = \boxed{4\frac{4}{5}}$

8. $\frac{10}{x} = \frac{8}{7}$

$\frac{8x}{8} = \frac{70}{8}$ $x = \boxed{8\frac{4}{8}}$ SIMPLIFY! $x = \boxed{8\frac{1}{2}}$

Solve the following proportions; leave your answers as fractions or mixed numbers

1. $\frac{x}{4} = \frac{3}{7}$

$x = \boxed{1\frac{5}{7}}$

6. $\frac{x}{7} = \frac{4}{5}$

$x = \boxed{5\frac{3}{5}}$

2. $\frac{n}{3} = \frac{5}{8}$

$x = \boxed{1\frac{7}{8}}$

7. $\frac{6}{5} = \frac{x}{9}$

$x = \boxed{10\frac{4}{5}}$

3. $\frac{2}{x} = \frac{3}{4}$

$x = \boxed{2\frac{2}{3}}$

8. $\frac{x}{100} = \frac{3}{4}$

$x = \boxed{75}$

4. $\frac{3}{5} = \frac{x}{6}$

$x = \boxed{3\frac{3}{5}}$

9. $\frac{r}{6} = \frac{7}{8}$

SIMPLIFY! $x = \boxed{5\frac{1}{4}}$

5. $\frac{8}{3} = \frac{x}{4}$

$x = \boxed{10\frac{2}{3}}$

10. $\frac{x}{8} = \frac{6}{7}$

$x = \boxed{6\frac{6}{7}}$

Intro to the Distributive Property

Example

Double burgers and double fries

×2   ×2

2(burgers + fries) = 2 burgers + 2 fries

Copy the + or - signs inside the parentheses

the burgers are doubled and the fries are doubled

DISTRIBUTING NUMBER

MULTIPLIES EVERYTHING INSIDE THE PARENTHESES!

Multiply everything inside the parentheses by the outside distributing number

1. 4(sandwich + drink)

   4 sandwiches + **4** drinks

2. 5(book + cover)

   5 books + **5** covers

   REMEMBER THE "BURGER-FRIES" TRICK!

3. 2(left shoe + right shoe)

   2 left shoes + **2 right shoes**

4. 3(salad + sandwiches + dessert)

   3 salads + **3 sandwiches** + **3 desserts**

© Peter Wise, 2014

56

Intro to the Distributive Property

Solve by multiplying everything inside the parentheses by the outside distributing number

1. (desk + lamp)2

   THE DISTRIBUTING NUMBER WORKS THE SAME AT THE END AS IT DOES AT THE FRONT!

   2 desks + **2** lamps

2. 3(dollar - penny)

   3 dollars - **3** pennies = **$2.97**

   Calculate the amount

3. 4(dozen - 2)

   4 dozen (48) - **8** = **40**

4. 5(dime + nickel + penny)

   50¢ + **25¢** + **5¢** = **80¢**

   Calculate amounts in cents

5. (quarter + dime + nickel)3

   $0.75 + **$0.30** + **$0.15** = **$1.20**

   Calculate amounts in dollars

© Peter Wise, 2014

57

Practice with the Distributive Property

Multiply everything inside the parentheses by the outside distributing number

Calculate Money Amounts

1. 4($2 - 1¢)

   $8 - **$0.04** = **$7.96**

   DID YOU REMEMBER TO MULTIPLY *BOTH* INSIDE NUMBERS BY 4?

2. 6(dime + nickel)

   $0.60 + **$0.30** = **$0.90**

3. (quarter + dime + penny)2

   $0.50 + **$0.20** + **$0.02**

   = **$0.72**

4. 3($10 bill + $5 bill + $1 bill)

   $30 + **$15** + **$3**

   = **$48**

Calculate Number Amounts

5. 3(20 + 5)

   60 + **15** = **75**

6. (12 + 5)4

   48 + **20** = **68**

7. 5(6 + 10)

   30 + **50** = **80**

8. 7(8 + 9)

   56 + **63** = **119**

© Peter Wise, 2014

58

Practice with the Distributive Property

Example

A. ×2   ×2

   2(10 + 7)   Copy the + or - signs inside the parentheses   2(10 + 7)

   2 · 10 + **2 ·** 7   ·2

   20 + **14** = **34**

Multiply everything inside the parentheses by the outside distributing number

1. 3(10 + 4)   3(10 + 4)

   3 10 + **3** 4   ×3

   30 + **12** = **42**

2. 2(100 + 1)

   2 100 + **2** 1

   200 + **2** = **202**

3. 2(100 - 1)   THIS IS A GREAT WAY TO MULTIPLY 2 X 99!

   200 - **2** = **198**

4. (10 + 4)7

   70 + **28**

   = **98**

5. (12 + 3)5

   60 + **15**

   = **75**

6. 3(1 dollar - 2 cents)

   $3 - **$0.06** = **$2.94**

© Peter Wise, 2014

59

Examples

the 2 multiplies everything in the parentheses

A. $2(a + b)$

$= \boxed{2a + 2b}$

the a multiplies everything in the parentheses

B. $a(c + d)$

$= \boxed{ac + ad}$

WHEN THEY TOUCH, THEY TIMES!

Expand, using the Distributive Property

Note that you are just rewriting these expressions, not finding an "answer"

1. $3(b + c)$

$3 \boxed{b} + 3 \boxed{c}$

2. $7(x + y + z)$

$\boxed{7} x + \boxed{7} y + \boxed{7} z$

3. $.2(a + b)$

$\boxed{.2a} + \boxed{.2b}$

4. $\frac{1}{2}(m + r)$

$\boxed{\frac{1}{2} m} + \boxed{\frac{1}{2} r}$

5. $5c(c + d)$

$\boxed{5c^2} + \boxed{5cd}$

WATCH CAREFULLY FOR EXPONENTS!

6. $(r + s)r^2$

THE DISTRIBUTING NUMBER CAN BE IN FRONT OR IN BACK—IT WORKS THE SAME EITHER WAY!

$\boxed{r^3} + \boxed{rs}$

7. $a^3(a^2 + ab - 2)$

$\boxed{a^5 + a^4b - 2a^3}$

8. $x^2y^3(x^5 - y^3)$

$\boxed{x^7y^3 - x^2y^6}$

© Peter Wise, 2014

60

Examples

$a \cdot a = a^2 \qquad a \cdot b = ab$

WHEN THEY TOUCH, THEY TIMES!

$a \cdot 2 = 2a$

A. $a(a + b)$

$= \boxed{a^2 + ab}$

B. $x(2 + y)$

$= \boxed{2x + xy}$

IT IS COMMON TO PUT THE LETTER AFTER THE NUMBER!

Expand, using the Distributive Property

Note that you are just rewriting these expressions, not finding an "answer"

1. $m(m + n)$

$\boxed{m^2} + \boxed{mn}$

2. $n(n - p)$

$\boxed{n^2} - \boxed{np}$

3. $r(2r + s)$

$2 \boxed{r^2} \boxed{+} \boxed{rs}$

PUT THE CORRECT SIGN HERE!

4. $y(y + 3)$

$\boxed{y^2} \boxed{+} \boxed{3y}$

5. $(x + y)x$

$\boxed{x^2} \boxed{+} \boxed{xy}$

6. $(m - n)n$

$\boxed{mn} \boxed{-} \boxed{n^2}$

7. $s(r + s - t)$

$\boxed{sr} \boxed{+} \boxed{s^2} \boxed{-} \boxed{st}$

8. $x(4 - x + y)$

$\boxed{4x} \boxed{-} \boxed{x^2} \boxed{+} \boxed{xy}$

© Peter Wise, 2014

61

Examples

A NEGATIVE DISTRIBUTING NUMBER OR LETTER REVERSES ALL THE SIGNS INSIDE PARENTHESES!

A. $-x(x + y - z)$

$-x^2 - xy + xz$

A NEGATIVE SIGN BY ITSELF OUTSIDE THE PARENTHESES IS REALLY (-1)!

B. $-(x + y - z)$

$-x - y + z$

IN THIS CASE, THE SIGNS JUST SWITCH, BUT EVERYTHING ELSE STAYS THE SAME!

C. $20 - (2 + 3 + 4)$

$20 \quad -2 \quad -3 \quad -4$

$= 20 - 9 = 11$

Anything negative outside the parentheses is a sign-switcher

Expand, using the Distributive Property

1. $-(a - b + c - d)$

IT SWITCHES THE SIGNS OF ALL THE NUMBERS INSIDE THE PARENTHESES!

$\boxed{-a + b - c + d}$

THE NEGATIVE NUMBER OUTSIDE IS A SIGN SWITCHER!

2. $-x(x - y - z)$

$\boxed{-x^2 + xy + xz}$

3. $-2a(2a + 3b - 4c)$

$\boxed{-4a^2 - 6ab + 8ac}$

4. $-10x^2(2x^3 - 3y^2)$

$\boxed{-20x^5 + 30x^2y^2}$

5. $-6m(-3m + 2m^2)$

$\boxed{18m^2 - 12m^3}$

6. $-4g^2h^3(5h - 3g^5h^2)$

$\boxed{-20g^2h^4 + 12g^7h^5}$

7. $-7r^4s^5(2r^2s^6 - 3r^5s^8)$

$\boxed{-14r^6s^{11} + 21r^9s^{13}}$

8. $-5m(5m - 7 - 6mn^3)$

$\boxed{-25m^2 + 35m + 30m^2n^3}$

© Peter Wise, 2014

62

Expand, using the Distributive Property

1. $x(x + y)$ — $\boxed{x^2 + xy}$

2. $x(x^2 + y^2)$ — $\boxed{x^3 + xy^2}$

3. $5c(c - d)$ — $\boxed{5c^2 - 5cd}$

Write the addition or subtraction signs

4. $(f + g)7f^3$ — $\boxed{7f^4 + 7gf^3}$

REMINDER! A NEGATIVE DISTRIBUTING NUMBER ACTS AS A SIGN SWITCHER FOR THE VALUES INSIDE THE PARENTHESES!

5. $-r(3 + s - 10r)$ — $\boxed{-3r - rs + 10r^2}$

A NEGATIVE SIGN IS THE SAME AS (-1) MULTIPLYING A NUMBER OR LETTER!

6. $4(x + y)$ — $\boxed{4x + 4y}$

7. $2(3n - m)$ — $\boxed{6n - 2m}$

8. $b^2(a + b)$ — $\boxed{b^2a + b^3}$

9. $7x(x + y)$ — $\boxed{7x^2 + 7xy}$

10. $-2a(4 - 3a)$ — $\boxed{-8a + 6a^2}$

11. $5r^2(12 + r^5)$ — $\boxed{60r^2 + 5r^7}$

12. $(9m + 7)4n^3$ — $\boxed{36mn^3 + 28n^3}$

© Peter Wise, 2014

63

Practice with the Distributive Property

Expand, using the Distributive Property

1. $6s^3(5 - 7s)$ $\boxed{30s^3 - 42s^4}$

2. $3y^4(9 - 6y^4)$ $\boxed{27y^4 - 18y^8}$

3. $3c^3(5 - 7s)$ $\boxed{15c^3 - 21c^3s}$

4. $-10c^6(-5d^2 + 4c)$ $\boxed{50c^6d^2 - 40c^7}$

5. $9g^5(3g^7 - 2h^4)$ $\boxed{27g^{12} - 18g^5h^4}$

6. $(3x^4 - 5y^3)7x^2$ $\boxed{21x^6 - 35x^2y^3}$

7. $8(7x - 3y)$ $\boxed{56x - 24y}$

8. $(4a + 2b - 3c)9$ $\boxed{36a + 18b - 27c}$

9. $12m^5(4 - 7m^2)$ $\boxed{48m^5 - 84m^7}$

10. $7r(7r + 4r^2)$ $\boxed{49r^2 + 28r^3}$

11. $-4x^7(3x^2 - 8y^5)$ $\boxed{-12x^9 + 32x^7y^5}$

12. $12n^4(2m^3 - 9n^2)$ $\boxed{24n^4m^3 - 108n^6}$

HINT: DISTRIBUTING NUMBERS HAVE THE SAME EFFECT WHETHER THEY'RE IN BACK OR FRONT OF PARENTHESES!

© Peter Wise, 2014

64

Distributive Property & Combining Like Terms

Expand, using the Distributive Property, then combine like terms

1. $5(3a + b) + 2(a + 4b)$

   #1 Expand, using the Distributive Property:

     $\boxed{15a + 5b}$ + $\boxed{2a + 8b}$

     $5(3a + b)$ $2(a + 4b)$

   #2 Combine like terms: $\boxed{17}$ a + $\boxed{13}$ b

2. $3(x + y) + 4(x + y)$

   #1 Expand, using the Distributive Property:

     $\boxed{3x + 3y}$ + $\boxed{4x + 4y}$

   #2 Combine like terms: $\boxed{7}$ x + $\boxed{7}$ y

3. $7(2m + 3r) + 3(2m + 10r)$

   #1 Expand, using the Distributive Property: $\boxed{14m + 21r}$ + $\boxed{6m + 30r}$

   #2 Combine like terms: $\boxed{20m}$ + $\boxed{51r}$

4. $3(2x - 5y) + 2(9y - 12x)$

   #1 Expand, using the Distributive Property: $\boxed{6x - 15y}$ + $\boxed{18y - 24x}$

   #2 Combine like terms: $\boxed{-18x}$ + $\boxed{3y}$

© Peter Wise, 2014

65

The Distributive Property with Equations

Example

A. $2(x + 3) = 16$

$2x + 6 = 16$ Expand, using the Distributive Property

$2x = 10$ Subtract 6 from both sides

$x = 5$ Divide both sides by 2

Expand, using the Distributive Property, then solve the equations

1. $2(x + 4) = 14$ x = $\boxed{3}$

   $2x + 8 = 14$ ← Expand, using the Dist. Prop.

   $2x = 6$ ← Subtract from both sides

   $x = 3$ ← Divide both sides

2. $3(x + 2) = 24$ x = $\boxed{6}$

   $3x + 6 = 24$ ← Expand, using the Dist. Prop.

   $3x = 18$ ← Subtract from both sides

   $x = 6$ ← Divide both sides

3. $2(x - 5) = 6$ x = $\boxed{8}$

   $2x - 10 = 6$

   $2x = 16$

   $x = 8$

4. $6(x - 2) = 48$ x = $\boxed{10}$

   $6x - 12 = 48$

   $6x = 60$

   $x = 10$

5. $-2(x + 3) = -14$ x = $\boxed{4}$

   $-2x - 6 = -14$

   $-2x = -8$

   $x = 4$

6. $-3(x - 4) = -15$ x = $\boxed{9}$

   $-3x + 12 = -15$

   $-3x = -27$

   $x = 9$

© Peter Wise, 2014

66

Algebra With Perimeters

Example

A. The perimeter of the rectangle is 36. What is the value of x?

   w = $3x + 1$

   $2(3x + 1) + 2(4x + 3) = 36$
     Double length   Double width

   Double this value for the width (or height)

   l = $4x + 3$

   $(6x + 2) + (8x + 6) = 36$

   Double this value for the length (base)

   $14x + 8 = 36$

   $14x = 28$ x = 2

PERIMETER MEASURES THE DISTANCE AROUND A 2-DIMENSIONAL SHAPE!

Perimeter of parallelograms = Double length + double width   2L + 2W!

1. The perimeter of the rectangle is 46 units. What is the value of a?

   You double each dimension for the perimeter of parallelograms

   $2x + 2$ [rectangle] $3x + 5$

   $\boxed{2} (\boxed{3x + 5}) + \boxed{2} (\boxed{2x + 2})$

   $\boxed{6x + 10}$ + $\boxed{4x + 4}$

   $\boxed{10x + 14}$ = 44 units   x = $\boxed{3}$

2. The perimeter of the rectangle is 80 cm. What is the value of y?

   $2y + 3$ [rectangle] $4y + 7$

   $\boxed{2} (\boxed{4y + 7}) + \boxed{2} (\boxed{2y + 3})$

   $\boxed{8y + 14}$ + $\boxed{4y + 6}$

   $\boxed{12y + 20}$ = 80 units

   x = $\boxed{5}$

© Peter Wise, 2014

67

Panel 1 (page 68)

Factoring: Reverse Distributive

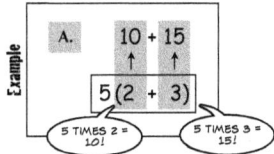

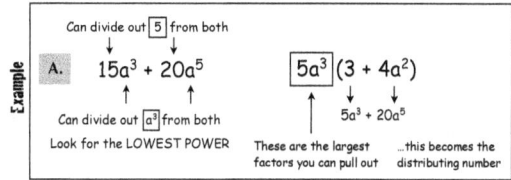

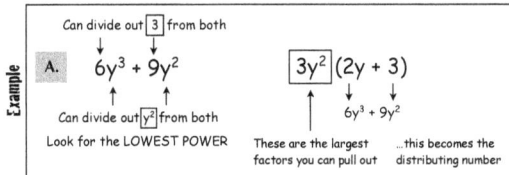

Example

A. $10 + 15$

$5(2 + 3)$

5 TIMES 2 = 10! 5 TIMES 3 = 15!

Factor, using the Distributive Property

1. $6 + 9$

$3(2 + 3)$

What do you multiply 3 by to get 6? What do you multiply 3 by to get 9?

5. $44 + 55$

$11(4 + 5)$

WHAT NUMBER DIVIDES EVENLY INTO 44 AND 55? What do you multiply the distributing number by to get 44? ...to get 55?

2. $21 - 28$

$7(3 - 4)$

6. $63 - 27$

$9(7 - 3)$

3. $40 + 60 + 80$

$20(2 + 3 + 4)$

7. $70 + 80$

$10(7 + 8)$

4. $30 + 54$

$6(5 + 9)$

8. $24 - 16 + 32$

$8(3 - 2 + 4)$

PUT THE CORRECT SIGNS BETWEEN THE BOXES!

68

Panel 2 (page 69)

Factoring: Reverse Distributive

Example

A. Can divide out 5 from both

$15a^3 + 20a^5$

Can divide out a^3 from both
Look for the LOWEST POWER

$5a^3(3 + 4a^2)$

$5a^3 + 20a^5$

These are the largest factors you can pull out ...this becomes the distributing number

Factor out the largest value from each to rewrite with the Distributive Property

With exponents...
Look for the lowest power

1. $6a^2 + 8a^3$

2 ← largest number that divides into both

a^2 ← largest power of a letter that divides into both... look for the LOWEST POWER!

$2a^2$ $(3 + 4a)$

THE NUMBER AND LETTER ABOVE FORM THE GCF. PUT BOTH OF THEM HERE FOR THE DISTRIBUTING TERM!

3. $14n^3 + 21n^5$

7 ← largest number that divides into both

n^3 ← largest power of a letter that divides into both...

$7n^3$ $(2 + 3 n^2)$

2. $12y + 15y^2$

3 ← largest number that divides into both

y or y^1 ← largest power of a letter that divides into both... look for the LOWEST POWER!

$3y$ $(4 + 5y)$

4. $30m^7 - 70m^{10}$

10 ← largest number that divides into both

m^7 ← largest power of a letter that divides into both...

$10m^7$ $(3 - 7 m^3)$

69

Panel 3 (page 70)

Factoring: Reverse Distributive

Example

A. Can divide out 3 from both

$6y^3 + 9y^2$

Can divide out y^2 from both
Look for the LOWEST POWER

$3y^2(2y + 3)$

$6y^3 + 9y^2$

These are the largest factors you can pull out ...this becomes the distributing number

Factor out the largest value from each to rewrite with the Distributive Property

1. $18a^4 + 27a^3$

9 ← largest number that divides into both

a^3 ← largest power of a letter that divides into both... look for the LOWEST POWER!

$9a^3$ $(2a^1 + 3)$

or 2a

3. $12m^{10} + 20m^7$

4 ← largest number that divides into both

m^7 ← largest power of a letter that divides into both... look for the LOWEST POWER!

$4m^7$ $(3m^3 + 5)$

2. $24c^5 + 36c^9$

12 ← largest number that divides into both

c^5 ← largest power of a letter that divides into both... look for the LOWEST POWER!

$12c^5$ $(2 + 3c^4)$

4. $21x^5 + 28x^2 - 14x^3$

7 ← largest number that divides into all three

x^2 ← largest power of a letter that divides into all three... look for the LOWEST POWER!

$7x^2$ $(3x^3 + 4 - 2x^1)$

70

Panel 4 (page 71)

Factoring Practice

Factor out the largest value from each to rewrite with the Distributive Property

1. $12r^5 - 8r^2$

4 ← largest number that divides into both

r^2 ← largest power of a letter that divides into both... look for the LOWEST POWER!

$4r^2$ $(3r^3 - 2)$

2. $5c^5 + 15c^6$

$5c^5$ $(1 + 3c^1)$

or 3c

largest number AND power of a letter that divide into both

3. $24s^7 + 27s^{10} - 18s^9$

$3s^7$ $(8 + 9s^3 - 6s^2)$

largest number AND power of a letter that divide into both

4. $63m^4n^3 + 45m^2n^5$

9 ← largest number that divides into both

m^2 ← largest power of m that divides into both... look for the LOWEST POWER!

n^3 ← largest power of n that divides into both... look for the LOWEST POWER!

$9m^2n^3$ $(7m^2 - 5n^2)$

5. $33a^2b^{11} - 77a^7b^3$

$11a^2b^3$ $(3b^8 - 7a^5)$

6. $20x^6y^{10} - 28x^4y^{15}$

$4x^4y^{10}$ $(5x^2 - 7y^5)$

Challenge Question:

7. $-3a^2 - 6y^2 - 5x^2$

$-($ $3a^2 + 6y^2 + 5x^2$ $)$

or -1

Hint: Look for the sign-switcher

71

Letters on One Side, Numbers on the Other

It actually doesn't matter which side you put the letters or numbers on

Simplest way: Look for the side with the greatest value for the letter.

Put the letter(s) on that side and the number(s) on the other side

Solve by getting the letters on one side and the numbers on the other side

1.

Letters ("variables") this side   Numbers ("constants") this side

$9x - 10 = 4x + 5$ $x = \boxed{3}$

The larger value of x is on this side. It's a little simpler to get the letter(s) on this side.

$5x - 10 = 5$ ← Subtract 4x from both sides (gets the variable out of the right side)

$5x = 15$ ← Add 10 to both sides (gets rid of the number on the left side)

$\dfrac{5x}{5} = \dfrac{15}{5}$ ← Divide both sides by the number multiplying the letter (the "coefficient")

2.

Constants this side   Variables this side

$-3x + 9 = 4x - 5$ $x = \boxed{2}$

The larger value of x is on the RIGHT side. It's a little simpler to get the letter(s) on this side.

$9 = 7x - 5$ ← Add 3x to both sides (gets the variable out of the left side)

$14 = 7x$ ← Add 5 to both sides (gets rid of the negative number on the right side)

$\dfrac{14}{7} = \dfrac{7x}{7}$ ← Divide both sides by the number multiplying the letter (the "coefficient")

3.

$2x + 2 = -2x + 30$ $x = \boxed{7}$

$4x + 2 = 30$ ← Add 2x to both sides (gets the variable out of the left side)

$4x = 28$ ← Subtract 2 from both sides (gets rid of the number on the left side)

$\dfrac{4x}{4} = \dfrac{28}{4}$ ← Divide both sides by the coefficient

72

4.

Numbers ("constants") this side   Letters this side

$4x - 4 = 7x - 28$ $x = \boxed{8}$

$-4 = 3x - 28$ ← Subtract 4x from both sides (gets the variable out of the left side)

$24 = 3x$ ← Add 28 to both sides (gets rid of the number on the right side)

$\dfrac{24}{3} = \dfrac{3x}{3}$ ← Divide both sides by the number multiplying the letter (the "coefficient")

5.

Letters this side   Constants this side

$16x - 5 = 2x + 40 + 5x$ $x = \boxed{5}$

$16x - 5 = 7x + 40$ ← Combine the x-values on the right side

$9x - 5 = 40$ ← Subtract 7x from both sides (gets the variable out of the right side)

$9x = 45$ ← Add 5 to both sides (gets rid of the negative number on the left side)

$\dfrac{9x}{9} = \dfrac{45}{9}$ ← Divide both sides by the number multiplying the letter (the "coefficient")

6.

$2x + 4x + 5x - 4 = 3x + 28$ $x = \boxed{4}$

$11x - 4 = 3x + 28$ ← Combine the x-values on the left side

$8x - 4 = 28$ ← Subtract 3x from both sides (gets the variable out of the right side)

$8x = 32$ ← Add 4 to both sides (gets rid of the constant on the left side)

$\dfrac{8x}{8} = \dfrac{32}{8}$ ← Divide both sides by the number multiplying the letter (the "coefficient")

7.

$12x - 22 = 3x + 5$ $x = \boxed{3}$

$9x - 22 = 5$ ← Subtract 3x from both sides (gets the variable out of the right side)

$9x = 29$ ← Add 22 to both sides (gets rid of the constant on the left side)

$\dfrac{9x}{9} = \dfrac{27}{9}$ ← Divide both sides by the number multiplying the letter (the "coefficient")

73

Multi-Step Equation Practice

Solve by getting the letters on one side and the numbers on the other side

1.

Variable this side   Numbers this side

$4(3x + 2) - 1 = -29$ $x = \boxed{-3}$

$12x + 8 - 1 = -29$ ← Expand, using the Distributive Property

$12x + 7 = -29$ ← Combine the like terms on the left side

$12x = -36$ ← Eliminate numbers on the left side

$x = -3$ ← Divide by the coefficient

2.

Variable this side   Numbers this side

$5x + 2(5 - 3) = -4 + 8$ $x = \boxed{0}$

$5x + 10 - 6 = -4 + 8$ ← Expand, using the Distributive Property

$5x + 4 = 4$ ← Combine the like terms on the left side

$5x = 0$ ← Get the numbers all on the right side

$x = 0$ ← Divide by the coefficient

3.

Numbers this side   Variable this side

$14x - 10 - 2 = 4(4x - 2) + 22$ $x = \boxed{-13}$

$14x - 10 - 2 = 16x - 8 + 22$ ← Expand, using the Distributive Property

$14x - 12 = 16x + 14$ ← Combine the constants on both sides

$-12 = 2x + 14$ ← Get the variable only on the right side (because the value of x is larger on that side)

$-26 = 2x$ ← Eliminate the constant on the right side

$-13 = x$ ← Divide by the coefficient

74

Equations: Variables on Both Sides

1. $4x = 32 + 2x$

$x = \boxed{16}$

2. $-7x = 36 - 11x$

$x = \boxed{9}$

3. $2x - 8 = 40 - 4x$

$x = \boxed{8}$

4. $4x - 70 = -3x + 7$

$x = \boxed{11}$

5. $-2x - 3 = -11x + 60$

$x = \boxed{7}$

6. $8x - 7x - 40 = -6x + 2$

$x = \boxed{6}$

7. $4x + 13 = -5 - 2x$

$x = \boxed{-3}$

8. $3x - 4 = -2x + 60 - 64$

$x = \boxed{0}$

9. $-3x + 30 = -10x - 5$

$x = \boxed{-5}$

10. $5x - 40 = 8 - 7x$

$x = \boxed{4}$

75

1. If a letter has no number in front of it, the invisible number is really $\boxed{1}$

2. Terms that have matching letters and matching exponents are called this (Example: $4x^2y^3$ and $5x^2y^3$) $\boxed{\textbf{like terms}}$

3. Write the equation for the following:

   A number doubled and increased by 2 equals the same number tripled and decreased by 7 $\boxed{\textbf{2x + 2 = 3x - 7}}$

4. $35°$ Give the value for x $\boxed{6}$ °

   $8x - 13°$

5. Use substitution to solve:

   $b^2 - 4ac = \boxed{17}$

   $a = 2 \quad b = 9 \quad c = 8$

6. Use substitution to solve:

   $(x^2 + 2x)mn = \boxed{120}$

   $x = 3 \quad m = 2 \quad n = 4$

   Circle the like terms
7. $\boxed{14y^3x^4} \quad 14x^3y^4 \quad \boxed{14x^4y^3} \quad \boxed{3y^3x^4}$

© Peter Wise, 2014

76

Simplify problems 8-12

8. $9x - 5x + 7x = \boxed{\textbf{11x}}$

9. $a^2b^5 \cdot a^4b^2c^3 = \boxed{\textbf{a}^6\textbf{b}^7\textbf{c}^3}$

10. $5a^3b^6 \cdot 7a^2b^5 \cdot b = \boxed{\textbf{35a}^5\textbf{b}^{12}}$

11. $12x^7y^8 \div 4x^3y^3 = \boxed{\textbf{3x}^4\textbf{y}^5}$

12. $\dfrac{5y}{x} + \dfrac{9y}{x} - \dfrac{2y}{x} = \boxed{\dfrac{\textbf{12y}}{\textbf{x}}}$

13. Your goal in solving equations is to $\underline{\textbf{isolate}}$ the variable

14. $8x + 18 = 74 \qquad x = \boxed{7}$

    $8x = 56$

    $x = 7$

    Show your steps!

15. $13 + 15 = 22x - 15x \qquad x = \boxed{4}$

    $28 = 7x$

    $4 = x$

    Show your steps!

16. $\dfrac{x}{9} = 6 \qquad x = \boxed{54}$

© Peter Wise, 2014

77

17. $-10x + 2x - 6x + 5x = \boxed{\textbf{11x}}$

18. Exponents indicate the number of repeated $\underline{\textbf{factors}}$

19. $3x = -9x + 48 \qquad x = \boxed{4}$

    $\dfrac{12x}{12} = \dfrac{48}{12}$

    Show your steps!

20. $\dfrac{3}{7}x = 12 \qquad x = \boxed{28}$

21. Solve the following proportion. Leave your answer as a mixed number in simplest form $\dfrac{3}{8} = \dfrac{x}{12} \qquad x = \boxed{4\frac{1}{2}}$

22. $9(a + b - c) = \boxed{\textbf{9a + 9b - 9c}}$

23. $(r^3 - s^4 + t^2)s^2 = \boxed{\textbf{r}^3\textbf{s}^2 - \textbf{s}^6 + \textbf{t}^2\textbf{s}^2}$

24. $-2b^3(7a^5 - 3b^5 + 4c^2) = \boxed{\textbf{-14b}^3\textbf{a}^5 + \textbf{6b}^8 - \textbf{8b}^3\textbf{c}^2}$

25. $8(x - 3) = 48 \qquad 8x - 24 = 48 \qquad \boxed{\textbf{x = 9}}$

    $8x = 72$

    $x = 9$

© Peter Wise, 2014

78

26. Factor (reverse Distributive Property)

    $21 - 35 \qquad \boxed{7}\left(\boxed{3} - \boxed{5}\right)$

27. Factor (reverse Distributive Property)

    FACTOR OUT THE LARGEST POSSIBLE FACTOR!

    $15y^2 + 20y^3 \qquad \boxed{\textbf{5y}^2}\left(\boxed{3} + \boxed{\textbf{4y}}\right)$

    Constants this side   Letters this side
28. $-3x + 9 = 4x - 5 \qquad x = \boxed{2}$

    $9 = 7x - 5$ ← Add 3x to both sides (gets the variable out of the left side)

    $14 = 7x$ ← Add 5 to both sides (gets rid of the negative number on the right side)

    $\dfrac{14}{7} = \dfrac{7x}{7}$ ← Divide both sides by the number multiplying the letter (the "coefficient")

29. $-2x - 4 = -10x + 60 \qquad x = \boxed{8}$

    $8x - 4 = 60$

    $\dfrac{8x}{8} = \dfrac{64}{8}$

30. $-6x - 13 = -15x + 50 \qquad x = \boxed{7}$

    $9x - 13 = 50$

    $\dfrac{9x}{9} = \dfrac{63}{9}$

© Peter Wise, 2014

79

Algebra Quiz

1. $y + y + y = $ **3y**

2. $(y)(y)(y) = $ **y^3**

3. $15c + 15c - 6c = $ **24c**

4. $3x^4 \cdot 7x^5 = $ **$21x^9$**

5. $5a^3b^6 \cdot 7a^2b^4 = $ **$35a^5b^{10}$**

6. $10a^9b^{12} \div 2a^2b^8 = $ **$5a^7b^4$**

7. Factor $18a^2 - 27a^3$ **$9a^2(2 + 3a)$**

   (Factor out as much as you can)

8. $2a^2 \cdot 2a^3 \cdot 3a^4 = $ **$12a^9$**

9. $2(x - 5) = 6$ $x = $ **8**

   2x - 10 = 6
   2x = 16
   ÷ 2   ÷ 2

   Show your steps!

10. $12x^5$ $\boxed{12x^3}$ $\boxed{4x^3}$

    Circle the like terms

11. Use substitution to solve:

    $4ac = $ **48** $a = 6$
    $c = 2$

    4(6)(2) = 48

12. Use substitution to solve:

    $2b^2 = $ **162** $b = 9$

    2(81) = 162

13. $8x + 12 = 44$ $x = $ **4**

    8x = 32
    ÷ 8   ÷ 8

    Show your steps!

14. $10x + 5x = 3x + 24$ $x = $ **2**

    15x = 3x + 24
    12x = 24
    ÷ 12   ÷ 12

    Show your steps!

15. $\frac{5}{9} = \frac{x}{3}$ $x = $ **$1\frac{2}{3}$**

    $\frac{9x}{9} = \frac{15}{9}$

    Express your answer
    as a mixed number
    in simplest form

16. $\frac{x}{5} = 7$ $x = $ **35**

Now that you've learned the basics of algebra, it's time to move on to the next level with

Linear Equations

Topics include:

- Slope
- Slope-Intercept Form
- Ordered Pairs in Different Quadrants
- Plugging in Values for "m" and "b"
- Canceling Fraction Slopes
- Solving Linear Equations by Making x/y Tables
- Plotting Line Points from an x/y Table
- Noting Differences in Linear Equations
- Finding Slope from Two Points
- Finding Slope from Rate of Change
- Finding the Equation of a Line from Slope and One Point
- Finding Intercepts
- Graphing When x or y Are Missing
- x and y as Solution Pairs
- Functions
- Finding Equations from the Coordinates of Two Points

...with more tips, tricks, and wackiness!

www.ingramcontent.com/pod-product-compliance
Lightning Source LLC
Chambersburg PA
CBHW051415200326
41520CB00023B/7245